Terrestrial Wireless Optical Communication

Terrestrial Wireless Optical Communication

Devi Chadha

Professor
Department of Electrical Engineering
Indian Institute of Technology, Delhi

New York   Chicago   San Francisco   Lisbon   London
Madri   Mexico City   Milan   New Delhi   San Juan
Seoul   Singapore   Sydney   Toronto

Cataloging-in-Publication Data is on file with the Library of Congress.

McGraw-Hill books are available at special quantity discounts to use as premiums and sales promotions, or for use in corporate training programs. To contact a representative please e-mail us at bulksales@mcgraw-hill.com.

Terrestrial Wireless Optical Communication

1 2 3 4 5 6 7 8 9 0   DOC/DOC   1 2 9 8 7 6 5 4 3

ISBN   978-0-07-181875-9
MHID   0-07-181875-8

The pages within this book were printed on acid-free paper.

To

*my inspiring and inquisitive grandson,
Varun*

Preface

There has been an unprecedented growth in the deployment of communication systems in last decade driven by the increasing demand of bandwidth and ubiquitous connectivity for data and other multimedia rich applications. Wireless communication is one of the most vibrant research areas for meeting these demands. The heterogeneous next generation communication systems are expected to have all hybrid technologies from satellite, fiber, copper, RF and Optical Wireless. Optical Wireless or Free Space Optics (FSO) has multi-gigabit bandwidth, and can provide cost effective and transparent network interoperability and expandability. FSO enables optical transmission speeds which are not possible using existing fixed wireless RF technology and obviates the need to buy expensive spectrum. This distinguishes FSO from fixed wireless technologies, such as IEEE 802.11, LMDS, and MMDS. This makes Optical Wireless a dynamic area of research and development in current times.

In this text the emphasis is to give a complete understanding of the subject area of Free Space Optical communication systems with both, basic concepts and design implementation in focus. It provides comprehensive coverage starting from atmospheric phenomena that affects light propagation, channel modelling and describes important issues in optical wireless theory, including diversity, coding and modulation and detection techniques. System design and analysis has been covered in sufficient detail to enable the designer to obtain performance parameters, link budget calculation and other design issues. The channel models, modulation and detection schemes are covered for, both, the indoor and outdoor FSO systems but the focus of the text is mainly on the terrestrial links. Though it is expected that readers of this book may have the background in optical fiber communication and communication systems engineering, it has been tried to introduce concepts from the first principles as much as possible. FSO is an emerging area for the short distance optical wireless networks and we hope this text will be used by the optical communication engineers, researchers, managers, and consultants to get the insight of this upcoming technology.

DEVI CHADHA

Acknowledgements

I would like to thank my Institute, Indian Institute of Technology (IIT), Delhi for granting me the sabbatical leave to start this project and providing the right stimulating environment to complete writing the book in time. I would also thank my faculty colleagues and the graduate students.

I would like to thank my Publisher, Tata McGraw-Hill for excellent editing and good interaction while compiling the final draft of the Book.

Foremost, I would like to thank my family for their patience, understanding and for their support.

Finally, I thank the Almighty God for giving me the strength to embark upon this project and with His blessings conclude it satisfactorily.

DEVI CHADHA

Contents

Preface   *vii*

Acknowledgements   *ix*

1.   Introduction—Optical Wireless Communication Systems   1

    1.1   Advantages and Disadvantages   3
    1.2   Classification   4
    1.3   Future Scope of FSO   11
    1.4   Structure of the Book   12
    References   13

2.   Wireless Optical Channels   15

    2.1   Free Space Optics Outdoor Channels   16
    2.2   Propagation in Terrestrial Link   16
    2.3   Wireless Optical Space Links   29
    2.4   Indoor Propagation Media   30
    Summary   32
    References   32
    Further Reading   33

3.   Channel Modeling   35

    3.1   Large and Small-Scale Signal Variation   36
    3.2   Input/Output Model of the Optical Intensity
           Wireless Channel   37
    3.3   Statistical Channel Modeling of FSO   43
    3.4   System Optical Transfer Function   53
    3.5   Indoor Channel Modeling   55
    3.6   Modeling Noise Sources in FSO   62
    Summary   62
    References   63

4.   Modulation Techniques   65

4.1   Constraints of the Channel   *66*
4.2   Types of Modulation Schemes   *66*
4.3   Selection Criteria   *67*
4.4   Vector Channel Modeling and Optimum Detection   *68*
4.5   Minimum Power and Spectral Bandwidth   *71*
4.6   Modulation Schemes Used in FSO Systems   *75*
Summary   93
References   94
Further Reading   95

**5.   Diversity and Detection Techniques in
          Optical Fading Channel   97**

5.1   Detection in an Optical Fading Channel   *98*
5.2   Signal Models for Detection   *102*
5.3   Detection in Single Input Single Output
      Turbulence Channels   *104*
5.4   Diversity   *107*
5.5   Spatial Diversity   *107*
5.6   MIMO Channel   *112*
Summary   125
References   126

6.   Channel Capacity   128

6.1   Channel Capacity of AWGN Channel   *129*
6.2   Capacity of Fading Channels   *134*
6.3   Channel Capacity of Single-Input
      Single-Output Atmospheric Optical Channel   *137*
6.4   Capacity of Optical Fading Channels with Diversity   *142*
6.5   Capacity of Photon Poisson Channels   *153*
References   158

7.   Coding in FSO Channels   160

7.1   Basic Concepts   *162*
7.2   Linear Block Codes   *164*
7.3   Tree Codes or Convolution Codes   *170*
7.4   Turbo Codes   *175*
7.5   Low Density Parity Check Codes   *181*
Summary   191
References   192

8. FSO Link and System Design   **194**

8.1   Link Design   *195*
8.2   Component Reliability   *203*
8.3   Eye Safety Consideration   *204*
8.4   Transceivers   *207*
8.5   Noise in FSO Receiver   *221*
Summary   *222*
References   *222*

Index   ***227***

Terrestrial Wireless Optical Communication

Introduction—Optical Wireless Communication Systems

Chapter 1

By definition, *Wireless Optical Communication* is the transmission of optical beams carrying information either through free space in the atmosphere or in confined environment. These systems work in the same IR range in which normally the optical fiber systems work. Historically, Optical Wireless (OW) or Free Space Optical (FSO) communication was first demonstrated by Alexander Graham Bell in the late 19th century. In the experiment, Bell converted sound waves into electrical telephone signals, and then transmitted them between the two transceivers through free air space along a beam of light for some 600 feet distance. Calling his experimental device the *photo phone*, Bell considered the optical wireless as the predominate invention and not the telephone, because it did not require wires for the transmission of the signal over the distance. Although Bell's photo phone never became a commercial reality, it demonstrated the basic principle of wireless optical communications.

The basic work on wireless optical communications systems was essentially started over the past 50 years or so, mostly for Defense and Space applications. The principal engineering challenges of wireless optics were addressed over the years through the activities in these areas. A basic optical wireless system, similar to fiber optical communications, consists of a transmitter using LEDs or LDs, but unlike fiber transmission, this unguided light beam modulated with analog or digital data containing video images, radio signals, or data files is collimated and transmitted through space rather than being guided through an optical fiber. These beams of light, operating in the IR wavelengths, are focused on receiving lens connected to a high-sensitivity receiver. Commercially available systems offer capacities in the range of 100 Mbps to 2.5 Gbps [1-3], and demonstration systems report data rates as high as 160 Gbps.

Unlike radio and microwave systems, free space optical communication requires no spectrum licensing and there is no interference to and from other systems. In addition, the point-to-point laser signal is extremely difficult to intercept, making it ideal for secure communications. Free space optical communications offer data rates comparable to fiber optical communications at a fraction of the deployment cost while extremely narrow laser beam widths provide no limit to the number of free space optical links that may be installed at a given location.

The advantages of free space optical wireless do not come without some cost. When light is transmitted through optical fiber, transmission integrity is quite predictable. On the other hand, unfortunately, when the light is transmitted through the air it must contend with a complex and not always predictable channel—*the atmosphere*. The fundamental limitation of free space optical communications arises from the environment through which it propagates. The free space optical communication systems though are relatively less affected by rain and snow, but can be severely affected by fog and atmospheric turbulence or scintillation. This can lead to decrease and fading in the power density of the transmitted beam, decreasing the effective distance and performance of the link. The detail description of the atmospheric effects on the IR transmission and its characterizations are given in the next chapter.

With the above brief introduction of free space optical communication systems, in section-1 of this chapter, we give in some more details the advantages and disadvantages of these systems over the fiber and the RF wireless systems. In section-2, we discuss the three types of OW communication systems, i.e., space, terrestrial and indoor. In section-3, we give the future role of FSO terrestrial communication systems in the ubiquitous communication scenario.

1.1   ADVANTAGES AND DISADVANTAGES

These systems offer:

➤ *Unlicensed spectrum:* The freedom of the IR spectrum from licensing and regulation translates into ease, speed and low cost of deployment. The only essential requirement for FSO or optical wireless transmission is Line of Sight (LOS) between the two ends of the link.

➤ *High speeds:* Speeds up to orders of Gbps using multi-beam and DWDM system are possible. Data rates comparable to optical fiber transmission can be carried by wireless optical systems with acceptable error rates.

➤ *Large bandwidth:* Broader bandwidth as compared to RF is possible as in optical fiber because of the higher optical carrier.

➤ *Low power requirement:* Much less consumption of power as compared to RF component counterparts.

➤ *No EMI:* Unlike radio and microwave systems FSO is an optical technology and hence no interference from or to other systems or equipment.

➤ *High security:* More secure as any interception is detected immediately. FSO laser beams cannot be detected with spectrum analyzers or RF meters. It requires a matching FSO transceiver carefully aligned to complete the transmission. Interception is, therefore, very difficult and extremely unlikely. Light beam cannot penetrate walls hence prevents eavesdropping.

➤ *Low-cost deployment:* Easier and cheaper installation as compared to deploying high-capacity fiber cables. Optical antennas and transceivers are much smaller in size, and hence, do not require much space to be installed as compared to RF antennas on the rooftops. In the last mile segment, it is also possible to mount these systems inside buildings as FSO transceivers can transmit and receive through the glass windows, reducing the need to compete for roof space, simplifying wiring and cabling and permitting FSO equipment to operate in a very favorable environment.

With all the above advantages over the RF and fiber transmission, FSO has its own drawbacks as well:

> ➢ Attenuation by the atmosphere and blocking by building/objects or personal as it is line of sight technology. Fading of the signal due to scintillation or turbulence in the atmosphere.

> ➢ Background noise of illumination of sun or lamps, etc., which are in the optical spectrum of the LED/LD.

> ➢ Link lengths in satellite links are several thousands of kilometers but terrestrial links are limited to few kilometers maximum because of the limit on the laser power transmission due to eye safety considerations.

1.2   CLASSIFICATION

Optical Wireless systems can be broadly classified as *Outdoor* and *Indoor* systems. The outdoor systems are popularly known as Free Space Optical systems. FSO links are point-to-point systems that transmit a modulated beam of visible or infrared light through the atmosphere. These are the *Space/Laser* links and the *Terrestrial* links. Indoor systems operate in limited and confined environment of a room. They are low cost, used for the connectivity of various devices and systems. The wavelength band used is from 780 nm to 950 nm. Therefore, Indoor systems are also known as IR systems.

On the basis of scope of application, OWS can be classified in three categories:

1.2.1   Space Communication

The space communication links are the air-to-ground, air-to-air space links, links between different earth orbits, i.e., LEO, GEO, etc., to ground, aircraft to ground or to other spacecraft, and deep-space or other planets to ground links. The advantages of Laser communication over traditional RF space communication systems are due to the narrower transmitted beam-width, orders of magnitude more bandwidth than RF, reduced payload weight due to smaller size equipment, and enhanced communication security due to the narrow beams. For example, in deep space missions, typical engineering considerations require inflight terminal optical antenna diameters of the order of 0.1 m to 0.3 m as compared to 1.5 m to 3 m RF antenna; a 10 m optical antennae on the ground compared with the 70 m RF dish antenna. This difference in sizes in turn results in advantages leading to increased power, reduced weight, and overall size reduction of the transceivers, which eventually is utilized to increase the data rate of the order of 100 or more of the FSO system over the RF system.

The advantage of optical communication derived from its comparatively narrow beam also introduces certain challenges. There is the difficulty of high-precision beam pointing. The RF beams require much less pointing precision; for example, from Mars, the beam footprint of X-band radar using a 3-m antenna is over 10,000 times the projected area of the Earth. At optical frequencies, the energy transmitted from a 30-cm antenna can be concentrated to about 1 per cent of the projected area of the Earth. Therefore, high precision is required to point the beam accurately to reliably hit the part of the Earth at the receiver. A sophisticated and complex system is used for acquisition, tracking and pointing (ATP) purposes [4].

Another challenge for optical communications is the difficulty of maintaining a communication link through cloud cover. The preferred optical communications connection with deep-space spacecraft would be through an optical receiver satellite, high above the clouds. There are several good references in laser space technology available in [5-7].

1.2.2 Free Space Terrestrial Links

The global telecommunications network has seen massive expansion over the last few years. First, there was tremendous growth of the optical fiber long-haul and wide-area network, followed by metropolitan-area networks. Deployment of the local-area networks and gigabit Ethernet ports also had similar growth rate. For this multi-gigabit network capacity to reach the information hungry end users, who always want to remain connected, what is required is a flexible and cost-effective means to access the broad array of services provided by the telecommunications network. However, most local loop network connections are not always broadband but limited to few Mbps [8]. As a consequence, there is a real need for a high-bandwidth bridge between the LANs and the MANs or WANs and make the bandwidth reach the end users. Free Space Optics systems can be one of the promising alternatives for addressing the emerging broadband access market and overcome the *last mile* bottleneck. Also, for the service providers installing FSO network or redeploying is cheaper and faster. Free Space Optics systems can function over distances of several kilometers as long as there is a clear line of sight between the two sites and enough power. Figure 1.1 illustrates the different types of connections possible with terrestrial links in a metro, local area and access network.

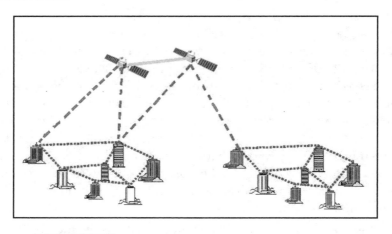

Figure 1.1   **Free Space Optical Terrestrial Links.**

Present access technologies are based on copper wire, broadband RF/microwave, coaxial cable and direct optical fiber connections by FTTC, FTTB or FTH. The telephone networks still use the Time Division Multiplex based network infrastructure which is limited to T-1 or E-1 and its increments. The broadband RF/microwave systems have the limitations of spectrum scarcity, congestion, and are expensive due to licensing. Radio equipment is also more expensive, the maximum data rates achievable with RF systems are low and the channels more insecure and subject to interference from and to other systems. Fiber deployment and re-deployment cost is very high and time consuming. Free space optical systems offer a broadband, flexible networking solution which is low in cost and more secure. FSO systems can take care of most of the limitations. It can with greater ease and speed bring the traffic to and from the optical fiber backbone. Another important factor which translates into ease, speed and low cost of deployment is freedom from licensing and regulation.

The challenges faced by the terrestrial links beside the LOS are the atmospheric attenuation and fading. Attenuation due to fog is very high at near-infrared wavelengths that are employed in FSO systems. Like any communication system, high reliability of the wireless optical system is essential, and therefore ways have to be engineered so that for a large fraction of the time an acceptable power will be received even in the presence of heavy fog. To avoid disruption either alternate redundant paths or dual diversity with RF/microwave transmission can be provided. An adaptive laser power scheme can be employed which can dynamically adjust the laser power according to weather conditions and thus, also improve the reliability of FSO systems. In clear weather the transmit power can be greatly reduced enhancing the laser lifetime while in low visibility conditions the laser power can be increased to maintain the optical link.

Performance of optical wireless systems is adversely affected by scintillation or turbulent atmosphere, which directly affects the bit error rate. To improve the performance in such conditions, various diversity schemes are used much like in RF wireless. Some optical wireless systems have a large aperture receiver, automatic gain control characteristics and clock recovery phase-lock-loop time constant which reduce the affects of atmospheric scintillation and jitter transference. Efficient coding schemes can also be incorporated to further improve on the performance. Further, the ambient light due to solar interference reduces the sensitivity of the system. This in FSO systems, operating at 1550 nm, can be reduced by using a long-pass optical filter to block all optical wavelengths below 850 nm from entering the system. Many of these issues will be discussed in detail in the book.

Another problem facing in FSO is the pointing stability because of building sway or tower movement. In the case of a robust fixed-pointed FSO system, which will be suitable for most deployments a combination of effective beam divergence and a well matched receiver Field of View (FOV) are provided. Fixed-pointed FSO systems are generally preferred over actively-tracked systems due to their lower cost.

1.2.3 Indoor Systems

These days, there are a large number of devices which need to interact with each other, such as, computers, printers, personal digital assistants, etc., used in home, offices and other establishments. These devices also require connection to the outside world for internet and web services. The wired interconnection network has the inherent problems of space, expandability, etc. They are also expensive and take a longer time to set up. To remove most of these bottlenecks, a wireless system is an attractive alternative. Both, radio frequency and infrared radiation, are possible options in implementing these wireless systems. The advantage of IR over RF has been discussed in detail in earlier sections. The IR indoor systems are similar to outdoor links. The wavelength band from 780 nm to 950 nm is the best choice for indoor optical wireless systems because low cost LEDs and LDs are readily available in this range. Low-capacitance silicon photodiodes also have their peak responsivity in this range, which increases the speed and sensitivity of the receiver.

The other advantage of IR signal is that it does not penetrate walls, and thus provides a security and privacy. For example, for a cell-based network in a building each room would be a cell and there would be no interference between the cells in different rooms, while all units of the different cells can be identical making it easier to keep the inventory. Indoor optical wireless systems have been used in many applications in the past few years, ranging from simple remote controls in home to more complex wireless local area networks. Several commercial products are now available and are being used successfully [9-12]. Due to the above reasons, optical wireless systems are becoming more popular in various operating environments, such as for consumer electronics in households, offices, medical facilities, and business establishments. Many other applications are envisaged for the future, including data networking in the indoor environment and the broadband multimedia services to mobile users within such an environment together with connectivity to base networks. There are certain limitations also in these systems, which have to be taken care of while designing the IR indoor system. Background ambient light induced by the artificial light sources falling on the detector surface produces noise [13]. Signals reaching through different paths cause multipath interference leading to inter-symbol-interference. There are also concerns of eye safety and power consumption which limit the average transmitter power.

The indoor system can have the following topologies in general:

Directed Beam or LOS Systems   In Directed Beam IR system (Fig. 1.2a), the optical beam travels directly without any reflection from the transmitter to the receiver. There is direct LOS link between the two fixed communication terminals with highly directional transmitter and receiver at both ends. Therefore, there is minimal path loss, higher power efficiency and higher transmission rates in these configuration. The main drawback of this technique is the lack of mobility, susceptibility to blocking by personnel and other objects and pointing problems due to narrow beams.

With the consideration of keeping the cost and operation ease one has to manually aim the transmitter towards the receiver unit. Keeping this in mind the beam-width should be chosen large enough [14].

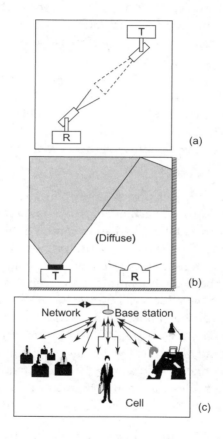

Figure 1.2 Indoor IR Configurations: (a) LOS [15] (b) Diffuse [17] (c) Cellular [17].

Diffused or Non-LOS Systems The diffused systems have lower transmission rates, increased robustness against shadowing and scope of high user mobility [15]. In Diffuse IR system (Fig. 1.2b), the transmitters send optical signals in a wide angle to the ceiling and after one or several reflections the signals arrive at the receivers. As the systems do not require a line-of-sight path for transmission, no alignment is required for setting up the connection. Nevertheless, the diffused systems have a higher path less than their LOS counterparts, requiring higher transmitter power levels and receivers with larger light collection area. The receiver, which collects beams traveling from different path lengths due to multiple reflections, suffers from multipath dispersion. This causes ISI at higher data rates or when the room size is larger in the cell system [16]. Therefore, in this configuration, the data rate depends on the room size and the reflection coefficients of the surfaces inside the room.

Cellular System   In cellular IR system (Fig. 1.2c) the base station is normally mounted on the ceiling of the room. Hot spots with minimum overlap are created by the transmitted beam, which can also provide a limited mobility. The base station has a relatively broad coverage and it transmits or receives the signal power to or from the remote terminals, respectively.

Wireless IR indoor LANs are an important alternative to wired and wireless RF LANs because of achievable high data rate, security and cost considerations [15-16]. Optical IR LANs can be either LOS or diffused. The diffuse Infrared LANs have almost uniform irradiance and, therefore, full mobility within the room is possible with no shadowing effects. Also, accurate alignment between transmitter and receiver will not be required. Different rooms can have identical transceivers for their particular LAN configuration and can be connected by wiring the base stations of the rooms by optical-fiber network. LOS IR LANs are commercially available [10]. JVC [12], IBM and Photonics [17-18] have modems that permit ad-hoc, peer-to-peer interconnection of notebook computers, which can handle bi-directional communications better than diffused LANs. These have higher data rates and better coverage but need terminal alignments.

Indoor IR Standards   Infrared Data Association (IrDA) has established standards [19-22] for short range, half-duplex line-of-sight (LOS) systems operating at bit rates up to 4 Mbps in 1993. Hewlett-Packard Company and IBM Corporation proposed Advanced Infrared (AIr) [23-24] standards in 1997. In addition in 1997, IEEE 802.11 standard [25] for wireless local area networks defines a specification for an infrared physical layer. The infrared physical layer has been designed for diffuse systems supporting two data rates (1 and 2 Mbps) and includes provisions for a smooth migration to higher data rates [26].

1.3 FUTURE SCOPE OF FSO

The next generation of wireless communication systems, the 4G systems will not be based on a single access technique, but it will have a number of different complementary access technologies to cater to the broadband multimedia requirements. It is envisaged that the future systems will not only connect users and their personal equipment but also have access to independent stand-alone equipment. Ultimately one would expect that everybody and everything will be wirelessly connected at all time and at all places. This scenario makes short-range communications more dominant and prevalent as the core network is already wired and available. In addition, these links will have to have high data throughputs with distances covering from a few kilometers down to sub-meter for WLANs, WPANs to WBANs, i.e., the wireless body area network. We see optical wireless communications to have a part to play in this wider 4G vision. The wireless optical channel has complementary characteristics to RF. It can in many situations, in order to improve the performance and capability of RF, add to and not replace RF in 4G wireless system. The FSO technology has matured now and is commercially viable to be used for establishing an optical backbone network, for extending and enhancing existing networks, and to be used in access segment. It offers broadband connectivity presently from 1.5 Mbps up to 4.0 Gbps with the advantage of being able to set up in short time and is a re-deployable solution. Both wireless and wireline carriers have tested and approved the technology for use in their networks, i.e., AT&T.

FSO has a future as a cost-effective connectivity alternative for several applications, for example, in microcell and cell-site backhaul for mobile networks, specifically for deploying 3G and high-density base stations. The primary bottleneck in a mobile network with 3G services is neither the microwave communication between a mobile device and a base station and nor the core network, it is the backhaul or link between the two. The base station has to transmit all the data being used by thousands of devices into one stream; hence the backhaul system has to have a very large bandwidth to provide good service. A solution to this problem can be with FSO. Another example can be as a carrier in geographies where climate makes microwave difficult, i.e., rain but no fog. Also where capital cost infrastructure is also a constraint. Similarly, in the enterprise network it can be used for LAN extension, private networks and last mile.

1.4 STRUCTURE OF THE BOOK

Optical wireless communication though encompasses all the three segments—the space, terrestrial and indoor—the respective channel characteristics are different, the application area performance requirements are different, and so on. Therefore, the system and technology used in each case varies. In this book, the terrestrial wireless communication is discussed with limited discussion of indoor and space communications.

In Chapter 2, the channel characteristics in the near-IR and IR range of the spectrum have been described. Understanding of the physical phenomenon associated with the channel is of great importance as it directly affects the designing of the transmitter and receiver systems of the FSO.

In Chapter 3, number of different models of atmospheric channel has been discussed. These mathematical models are for the fading and multipath characteristics of the channel. Towards the end, the model for the indoor wireless system is also described.

In Chapter 4, modulation techniques relevant to the FSO communication system are given. Not all modulation schemes used in RF are applicable to optical signal as they are intensity signal and so non-coherent detection is the method of choice. The performance of the various modulation schemes with the optical channel is discussed. In Chapter 5, the detection techniques used in these systems, both for single-input-single-output and multi-input-multi-output FSO systems are described.

In Chapter 6, we discuss the optimal performance achievable in a FSO channel. One of the basic measures of optimal performance is the capacity of a channel. In this, we also discuss the capacity enhancement methods in the terrestrial links. In Chapter 7, the different forward-error coding schemes in FSO systems are discussed for the performance improvement in these systems, which are affected by fading and noise.

In Chapter 8, the link and system design issues for the terrestrial links are discussed. The important issues of reliability and eye safety standards in these links are also given.

References

1. Lightpointe; www.lightpointe.com
2. PWComms;www.comms.co.uk
3. fSONA;www.fsona.com
4. David G. Aviv, *Laser Space Communications*, Artech House-NY, 2006
5. Ortiz G.G., Lee S., Alexander W. 'Sub-Micron Pointing System Design for Deep Space Optical Communication', SPIE Proc., Vol. 4272, 19th International Commun. Sat. System Conf. 2001.
6. Boroson D., Bondurant R.S. and Scozzafava J.J., 'Overview of High Rate Deep Space Laser Communications Options', SPIE Proceedings, Vol. 5338, Free Space Laser Communications Technologies XVI, ed., S. Mecherle, January 27, 2004.
7. Toyoshima M., 'Special Report: Trends of research and development of optical space communication technology', Space Japan Review 12-1, No. 44, 2005
8. Kedar D. andArnon S., 'Urban Optical Wireless Communication Networks: The main challenges and possible solutions', *IEEE Optical Communications*, May 2004, pp. 52–57.
9. Gfeller F.R. andBapst U.H., 'Wireless In-House Data Communication via Diffuse Infrared Radiation', *Proceedings of IEEE*, Vol. 67, No. 11, November 1979, pp. 1474–1486.
10. Pahlavan K., Probert T., and Chase M., 'Trends in Local Wireless Networks', *IEEE Commun. Mag.*, Mar. 1995, pp. 88–95.
11. Lessard et al., 'Wireless Communication in the Automated Factory Environment', *IEEE Network*, vol.2, no.3, May 1988, pp.64-69.
12. JVC VIPSLAN-10: www.jvcinfo.com.
13. Kahn J.M. and Barry J.R., 'Wireless Infrared Communications', *Proceedings of the IEEE*, vol. 85, no. 2, February 1997, pp. 265-298.
14. David J.T., Heatley, at. el, 'Optical Wireless: The Story So Far', *IEEE Commun. Magazine*, December 1998.
15. Barry J.R. et al. 'High speed-Non-Directive Optical Communication for Wireless Network', *IEEE Network*, Vol. 5, no. 6, Nov.1991, pp. 44-54.
16. Barry J.R. et al. 'Non-directed Infrared Links for High Capacity Wireless LANs', *IEEE Personal Commn*, second quarter1994, pp12-25.
17. IBM Corporation, Armonk, NY: www.ibm.com
18. Photonics Corporation San Jose, CA: www.Photonics.com
19. Infrared Data Association, Serial Infrared Physical Layer Specification, Version 1.3, October 15, 1998.
20. Infrared Data Association, Serial Infrared Physical Layer Link Specification, Version 1.1, October 1995.
21. Infrared Data Association, Serial Infrared Link Access Protocol (IrLAP), Version 1.1, June 1996.
22. Infrared Data Association, Link Management Protocol (IrLMP), Version 1.1, January 1996.

23. Hewlett-Packard Company and IBM Corporation, Request for Comments on Advanced Infrared (AIr) IrPHY Draft Physical Layer Specification, Version 0.4, January 1998.

24. Hewlett-Packard Company and IBM Corporation, Request for Comments on Advanced Infrared (AIr) IrMAC Draft Protocol Specification, Version 0.2, July 1997.

25. Project IEEE 802.11, IEEE Standard for Wireless LAN Medium Access Control (MAC) and Physical Layer (PHY) Specifications, Draft 6.1, June 1997.

26. R-Iniguez R. , Idrus S.M. and Sun Z. , *Optical Wireless Communication-IR for Wireless Connectivity*, CRC Press-2008.

Chapter 2

Wireless Optical Channels

In Chapter 1 we have discussed the two broad categories of wireless optical systems—the outdoor free space systems that gives both the space and terrestrial links, and the confined indoor systems. The optical signal experiences different propagating media in all the three cases and, therefore, the channel characteristics in these systems are not the same. In this chapter, we give the physical characteristic of the wireless optical channel, both in the outdoor and indoor environment. Understanding of the physical phenomenon associated with the channel and the deployment details of the link are of great importance when designing the transmitter and receiver systems of the FSO.

With a brief overview of the outdoor free space channel in Section 2.1, Section 2.2 discusses in detail the propagation characteristics of the terrestrial atmospheric link. In Section 2.3, we go on to discuss few more atmospheric conditions to be taken into account in the case of space optical links. Similarly, Section 2.4 discusses the indoor propagation media, which is quite different from the outdoor atmospheric conditions.

2.1   FREE SPACE OPTICS OUTDOOR CHANNELS

Free space optical communication involves the use of optical links across the space between two points, either terrestrial links within the Earth's atmosphere, or satellite links or links in outer space. Figure 2.1 shows a generic free space optics system. On the transmitting side, the transmitter consists of an optical antenna with other optical components, usually lens and tracking mechanics to direct the beam toward the receiver. On the receiving side, the receiver antenna also has lens to focus the received beam to the optical detector. For longer distances in a transceiver, a pair of telescopes is used at each end as antenna, with a laser and photo-sensors mounted in each telescope.

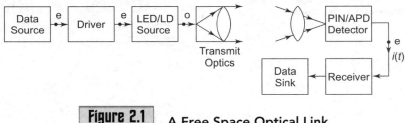

Figure 2.1   A Free Space Optical Link.

First, we discuss free space optics channels mainly for the application as a terrestrial communication link similar to Fig. 2.1 to serve as either the last mile or a LAN link. Usually, for reason explained later, wavelengths of 850 nm (short) and 1550 nm (long) are used for FSO systems.

2.2   PROPAGATION IN TERRESTRIAL LINK

While fiber-optic cable and FSO technology share many of the same attributes, they face different challenges due to the way they transmit information. Optical fiber is a very stable medium with minimal outside disturbances. FSO technology is subject to its own potential outside disturbances in the atmosphere. A laser beam propagating through the Earth's atmosphere is subjected to attenuation due to absorption of the light by atmospheric constituents and scattering by particles in the atmosphere. The shape, direction and electromagnetic properties of the laser beam are affected by these atmospheric disturbances and, therefore, the power budget and the overall performance of the free space optic link are strongly determined by them. Optical wireless networks based on FSO technology must be designed to combat the changes in the atmosphere which can affect FSO system performance.

The factors affecting the transmission of the laser beam in free space links are the:

➤ Beam divergence
➤ Atmospheric losses
➤ Atmospheric Turbulence
➤ Ambient light

2.2.1 Beam Divergence

At the transmitter, the beam divergence is caused by diffraction around the aperture at the end of the telescope as shown in Fig. 2.1. Divergence determines how much useful signal energy will be collected at the receive end of a communication link. It also determines how sensitive a link will be to displacement disturbances. Of the processes that cause attenuation, divergence is the only factor that is independent of the transmission medium; it will occur in vacuum just as much as in a stratified atmosphere. The full divergence angle, θ of the beam spread is approximately [1]:

$$\theta \approx \frac{\lambda}{D} \tag{2.1}$$

where, λ is the laser wavelength and D is the aperture diameter. The above rule is widely used to provide quickly an estimate of the size of the laser beam and can be used for atmospheric application as well. A better approximation of the angular spread, $\theta_{1/2}$ (half angle of the cone), of a beam projected along an atmospheric path [2] is:

$$\theta_{1/2} = \sqrt{\frac{1}{k^2 a^2} + \frac{1}{k^2 \rho^2}} \tag{2.2}$$

where, $k = \dfrac{2\pi}{\lambda}$, a is the beam radius when projected, ρ is the transverse coherence distance related to the effect of atmosphere turbulence on the propagation of light, as will be discussed later.

In practice, FSO-transmitted beam gets defocused from the diffraction limit so as to be larger than the diameter of the telescope at the receive end. This, indeed, helps to maintain alignment with the receiver in the face of random displacement disturbances. But, this also causes losses. Figure 2.2 shows the loss due to beam divergence. As an example, for

FSO with a range of 1 km the diameter of the divergent light beam is about 1 m for a very narrow beam diameter of 1 milliradian. When such a large diverged laser beam falls on the receiver lens, there is a significant loss due to the beam divergence, unless the collection aperture has a diameter larger than 1 m.

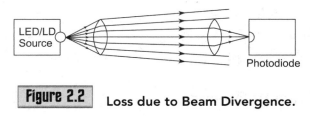

Figure 2.2   **Loss due to Beam Divergence.**

2.2.2   Atmospheric Losses

The physics of atmospheric losses or attenuation for the optical beam is attributed to *absorption* and *scattering* phenomenon in the medium. Absorption is the conversion of energy; reduction from the energy of photons in the light signal beam to the increase in the internal energy of the absorbing particle. Scattering is caused when the light collides with some particles or the scatterer and is deflected in arbitrary directions. Scattering can be categorized according to the physical size of the scatterer. When the scatterer is smaller than the wavelength, it is known as *Rayleigh scattering* with a very small scattering coefficient; when the scatterer is of comparable size to the wavelength, it is known as *Mie scattering*. When the scatterer is much larger than the wavelength, it is known as *Non-selective scattering*. There is no loss of energy in scattering; there is only a directional redistribution of energy to other angles. This may cause significant reduction in beam intensity for longer distances in the direction of propagation. The impact of absorption and scattering on the transmission of light through the atmosphere is discussed in more detail in the following sections.

Beer's Law   Beer's Law describes the attenuation of light traveling through the atmosphere due to both, absorption and scattering. In general, the transmission coefficient $T(z)$ of laser radiation in the atmosphere as a function of distance z, is given by *Beer's Law*, as [3]

$$T(z) = \frac{I_z}{I_0} = e^{-\gamma(\lambda)z} \tag{2.3}$$

where, I_z/I_0 is the ratio of the detected intensity I_z at the location z and the initially launched intensity I_0, and $\gamma(\lambda)$ is the attenuation coefficient, which is a function of the wavelength.

The attenuation coefficient is a sum of four individual parameters— molecular and aerosol *absorption coefficients* α_m and α_a, and molecular and aerosol *scattering coefficients* β_m and β_a, respectively, each of which is a function of the wavelength. The attenuation coefficient, therefore, can be expressed as:

$$\gamma = \alpha_m + \alpha_a + \beta_m + \beta_a \tag{2.4}$$

This formula shows that the total attenuation represented by the attenuation coefficient γ, results from the superposition of various scattering and absorption processes.

2.2.2.1 Absorption

The presence of the abundance of absorbing particles in the atmosphere determines how strongly the optical signal will be attenuated. The vibrational and rotational energy states of these particles are capable of absorbing optical beam in many bands. These particles can be divided into two general classes: *molecular* and *aerosol* absorbers.

Molecular absorption is due to the interaction of the laser beam with the gas molecules in the medium, viz., N_2, O_2, H_2, etc. The particles are characterized by their index of refraction. The imaginary part of the index of refraction, n_{im}, is related to the absorption coefficient, α_m, by the following equation [3]:

$$\alpha_m = \frac{4\pi n_{im}}{\lambda} = \sigma_{abs} N \tag{2.5}$$

where, σ_{abs} is the absorption cross section and N is the concentration of the absorbing particles. In other words, the absorption coefficient is a function of the absorption strength of a given type of particles, as well as a function of the particle density. Molecular absorption is wavelength depended and occurs more readily at some wavelengths than others. This selective spectral transmission in the atmosphere leads its transmission characteristics to have transparent and opaque windows similar to optical fibers. It gives rise to low-loss transmission windows centered on 850 nm, 1300 nm and 1550 nm as shown in Fig. 2.3. The low-loss windows coincide with those in fibers. This means that the same optoelectronic devices can be used in optical wireless systems as well.

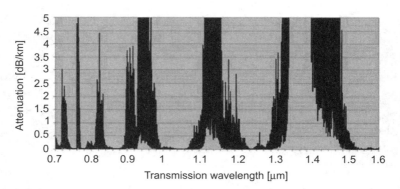

Figure 2.3 Transmission Windows of Optical Signal in Near Infrared Region Under Clear Weather Conditions. The calculation was done by MOTRAN [Source: AirFiber Inc.].

Aerosols are the suspended solid or liquid particles present in the atmosphere in the form of fog, mist, etc., which are liquid aerosols; dust, sea-salt particles, desert dust, and volcanic debris are the solid aerosols. They can also be created as a result of man-made chemical conversion of trace gases to solid and liquid particles and as industrial waste. These particles can range in size from fine dust less than 0.01 μm to large particles greater than 10.0 μm.

2.2.2.2 Scattering

Scattering of light can drastically impact the performance of FSO systems. There is a significant reduction of received light intensity at the receiver location due to the redirection or redistribution of light. Several scattering regions exist, depending on the characteristic size of the particles. The scattering regimes can be classified by size parameter Ω described by:

$$\Omega = \frac{2\pi r}{\lambda}$$

(2.6)

where, λ is the transmission wavelength and 'r' is the particle radius. For $\Omega \ll 1$, the scattering is in the Rayleigh regime; for Ω to be the order of 1, the scattering is in the Mie regime; and for $\Omega \gg 1$, the scattering is non-selective as the attenuation of the beam due to scattering is wavelength independent.

In the following paragraphs, we get some more details of these regimes.

Rayleigh scattering When the particle size is much smaller than the wavelength, Raleigh scattering is prominent. A radiation incident on the bound electrons of an atom or molecule of a gas induces a charge imbalance or dipole that oscillates at the frequency of the incident radiation. The oscillating electrons reradiate the light in the form of a scattered wave. Raleigh's classical formula for the scattering cross-section is as follows [4]:

$$\sigma_s = \frac{2\pi^5}{3}\left(\frac{n^2-1}{n^2+2}\right)^2\frac{(2r)^6}{\lambda^4} \tag{2.7}$$

where, 'n' is the refractive index . The λ^{-4} dependence in (2.7) implies that shorter wavelengths are scattered much more than longer wavelengths in atmosphere. Therefore, for FSO systems operating in the longer wavelength near infrared wavelength range, the impact of Rayleigh scattering on the transmission signal can be neglected. Though O_2 and N_2 do not cause absorption of IR signal, but they do participate in IR Rayleigh scattering. The wavelength dependence of the Rayleigh scattering cross-section in the infrared spectral range is shown in Fig. 2.4

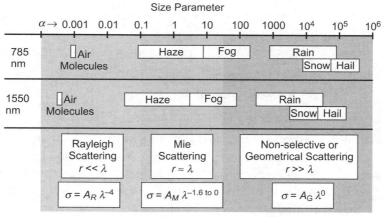

σ = Scattering coefficient, A_R, A_M, A_G = cnstants; λ = wavelength

Figure 2.4 Size Parameters and Corresponding Regions of Atmospheric Scattering Particles for 785 and 1500 nm [5].

Mie Scattering The Mie scattering region occurs for particles about the size of the wavelength. Therefore, as shown in Fig. 2.4, in the near infrared wavelength range, fog, haze, and pollution (aerosols) particles are the major contributors to the Mie scattering process. The theory is well understood, but there exists a problem in comparing the theory to an experiment. This is because the experimental data is collected in an atmospheric window with the assumption that only scattering is taking place, but it is the absorption that dominates most of the spectrum. In addition, the particle distributions is not known accurately. For aerosols, this distribution depends on location, time, relative humidity, wind velocity, and so on. An empirical simplified formula found in literature [6] used in the FSO system design for a long time to calculate the attenuation constant due to the Mie scattering is given by the following:

$$\gamma = \frac{3.91}{V}\left(\frac{\lambda}{550}\right)^{-q} \tag{2.8}$$

where γ is the atmospheric attenuation coefficient and q the size distribution of the scattering particles is:

q = 1.6 for high visibility (V> 50 km)

= 1.3 for average visibility (6km<V<50km)

= 0.585 $V^{1/3}$ for low visibility (V<6 km)

A more recent study [5] proposes another expression for q as:

= 1.6 for high visibility (V> 50 km)

= 1.3 for average visibility (6km<V<50km)

= 0.16V + 0.34 for haze visibility (1km<V<6km)

= V- 0.5 for mist visibility (0.5km<V<1km)

= 0 for fog visibility (V<0.5 km)

In (2.8), V corresponds to the visibility in Km, and λ is the transmission wavelength in nm. The transmission wavelength dependency of the attenuation coefficient γ does not follow the predicted empirical formula. More precise numerical simulations of the exact Mie scattering formula suggest that the attenuation coefficient does not drastically depend on wavelength in the near infrared wavelength range typically used in FSO systems is concerned.

2.2.2.3 Atmospheric Factors Affecting Transmission in FSO Links

In the following section, we discuss the atmospheric factors or the aerosols, which cause absorption and scattering in systems.

Fog The primary challenge to FSO-based communications is fog. This causes both absorption as well as scattering. As seen in Fig. 2.5, comparatively speaking, rain and snow have little effect on FSO systems than fog. Fog is vapor composed of water droplets, which are only a few hundred microns in diameter, but can modify light characteristics or completely hinder the passage of light through a combination of absorption, scattering, and reflection. The particle size distribution varies for different degrees of fog. Weather conditions are typically referred to as foggy when visibilities range between 0–2,000 m. Foggy conditions are many times described as *dense fog* or *thin fog* to characterize the appearance of fog. The density distribution of fog particles can also vary with height, which makes the modeling of fog more complex. Another factor to keep in mind is that the weather conditions are typically measured at airports which can be located away from the actual FSO installation location; hence the figures for the visibility may not be exact. The local impact of fog on the availability of FSO systems is one of the biggest challenges in the design of FSO links. Fog can be extremely thick, with attenuation values of 350 dB/km or more. Even modest fog conditions can highly attenuate infrared signals over shorter distances.

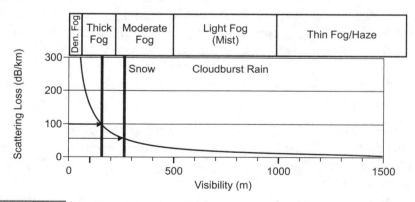

Figure 2.5 Attenuation as a Function of Visibility and the Weather Condition due to the Visibility [5].

Rain   Rain also causes attenuation in FSO systems, although its impact is significantly less than that of fog. This is because the radius of raindrops (200–2000 μm) is significantly larger than the wavelength of typical FSO light sources. Even in heavy rain, the absorption loss of water normally gives an attenuation loss of about 10 dB/km around 800 and 1500 nm, significant less than the scattering loss of fog. Extremely heavy rain can distrupt a link totally but it only occurs if one cannot see through the rain. Clouds are normally there along with rain or snow. If the clouds are very low to obstruct the laser beam, the attenuation can be as high as fog. Rain may also bring a water sheet on the glass of the windows. This may deflect part of the laser beam reaching the receiver if the transceiver is mounted inside the window pane. Typical rain attenuation values are moderate in nature. For example, for a rainfall of 2.5 cm/hour, a signal attenuation of 6 dB/km can be observed. Therefore, commercially available FSO systems that operate with a 25 dB link margin can penetrate rain relatively undisturbed.

Snow   Snow usually has a droplet size between fog and rain. There can be many shapes and sizes of snowflakes. The scattering and absorption loss of snow is between fog and rain. Heavy snow may cause ice build-up in the window panes and then can block the whole laser beam. In general, however, attenuation due to snow tends to be larger than rain. Whiteout conditions might attenuate the beam, but as the size of snowflakes is larger compared to the operating wavelength, scattering does not tend to be a big problem for FSO systems. The impact of light snow to blizzard and whiteout conditions fall approximately between light rains to moderate fog, with link attenuation potentials of approximately 3 – 30 dB/km.

Sand   Sand particles can scatter the optical beam and reduce the light intensity. In desert areas sand storm may also bring a link down.

   To summarize, although it is not possible to change the physics of the atmosphere, it is possible to take advantage of optimal atmospheric windows by choosing the transmission wavelengths accordingly. To ensure a minimum amount of signal attenuation from scattering and absorption, FSO systems operate in low-loss atmospheric windows in the IR spectral range. In the near infrared, water vapor is the primary molecular absorber with many absorption lines to attenuate the signal. The absorption due to oxygen and carbon dioxide are less and can be easily avoided. Figure 2.6 shows several transmission windows that are nearly transparent with an attenuation coefficient of less than 0.2 dB/km within the 0.8 to 1.6 micron wavelength. Majority of FSO systems are designed to operate in the windows of 780–850 nm and 1520–1600 nm. In the 1520–1600 nm window, 50–65 times more power can be transmitted than at 780–850 nm for the same eye safety classification, owing to the low transmission of the human eye at these wavelengths.

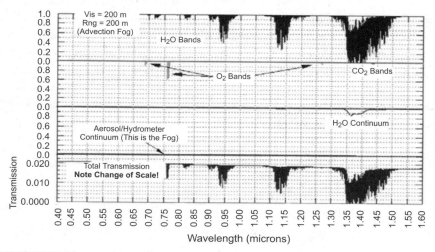

Figure 2.6   **Various Sources of Atmospheric Loss due to Absorption by Air and Fog Scattering. The Fog Visibility is 200 m and the Measured Range is also 200 m in Advection Fog. [Source: COLT Telecom].**

2.2.3 Atmospheric Turbulence

Turbulence is a random phenomenon, which occurs in the atmosphere due to the change in the refractive index of the air with temperature. Heated air rising from the earth surface creates temperature variations among the air pockets. The atmosphere acts like a collection of many prisms and lenses due to the index variation. These deflect the light beam into and out of the transmit path. Because these air pockets are not stable in time or in space, the change of index of refraction appears to follow a random motion. This appears as turbulent behavior in the atmosphere [7]. A good measure of turbulence is the refractive index structure coefficient C_n^2. This varies with time of the day and also with height. Because the air needs time to heat up, the turbulence is typically greatest in the middle of the afternoon ($C_n^2 = 10^{-13}$ m$^{-2/3}$) and weakest an hour after sunrise or sunset ($C_n^2 = 10^{-17}$ m$^{-2/3}$). C_n^2 is usually largest near the ground, decreasing with altitude.

The propagating laser beams experience three effects under atmospheric turbulence condition:

Scintillation Scintillation is the temporal and spatial fluctuations of the laser beam due to turbulence. Scintillation causes the optical signal to scatter preferentially at very shallow angles in the direction of propagation. These multiple signals phase shifted relative to each other arrive simultaneously at the receiver. This in turn causes the amplitude of the received signal to fluctuate randomly by as much as 30 dB if conditions are unfavorable. The power spectral density of these fluctuations typically spans 0.01–200 Hz, and hence can give rise to long bursts of data errors. Scintillation effects for small fluctuations follow a log-normal distribution, characterized by the Rytov variance, σ_R for a plane wave given by the following [3]:

$$\sigma_R^2 = 1.23 C_n^2 k^{7/6} L^{11/6} \tag{2.9}$$

where, $k = \dfrac{2\pi}{\lambda}$ and L as the link length. This expression suggests that larger wavelengths would experience a smaller variance for a given link. The variance in the case of spherical wavefront will be somewhat different. For FSO systems with a narrow, slightly diverging beam, the plane wave expression is more appropriate than that for a spherical beam. For the case, when the wavefront is curved when it reaches the detector, the transmitting beam being much larger than the detector the wavefront would be effectively flat. When the turbulence is high, the variance expression for large fluctuations is as follows:

$$\sigma_{high}^2 = 1.0 + 0.86 (\sigma_R^2)^{-2/5} \tag{2.10}$$

The above expression suggests that in case of larger fluctuation the shorter wavelengths would experience a smaller variance. For strong scintillation, the distribution tends to be more exponential.

In FSO deployment, the beam path must be more than 5 m above city streets or other potential sources of severe scintillation.

Beam Wander In the condition, when the turbulent eddies or the changing refractive index cells in the atmosphere become larger than the beam, the beam can be deflected randomly from its direction of propagation. This phenomenon is known as *beam wander* or *image dancing*. The light will be focused or defocused randomly at the detector following the index changes of the transmission path because of the refraction through the atmosphere, much as when the light passing through a refractive media such as a glass lens. This would, therefore, require larger than normal FOV of the photodetector ensuring that the signal is never lost. Beam wander is generally not significant over distances less than 500 m, but increases rapidly with distance. It is present both for a collimated beam on terrestrial horizontal path and for an uplink path in a space. In this case of beam wander for a beam in the presence of large cells of turbulence compared to the beam diameter, geometrical optics can be used to describe the radial variance, σ_r, as a function of wavelength and distance L, as follows:

$$\sigma_r^2 = 1.83 C_n^2 \lambda^{-1/6} L^{17/6} \tag{2.11}$$

From the above relationship, we observe that the wavelength dependence is weak as compared to scintillation effect. The longer wavelengths will be less effected by the beams wander than the shorter wavelengths. The rate of fluctuations is slow, less than a kHz or two.

Beam Spreading   The beam can spread more than the diffraction theory predicts due to turbulent atmosphere. Beam spreading, which can be both long-term and short-term, is the spread of an optical beam as it propagates through the atmosphere. The beam size can be characterized by the effective radius, a_{eff}, the distance from the center of the beam to where the relative mean intensity has decreased by $1/e$. The effective radius is given by the following [4]:

$$a_{eff} = 2.01 (\lambda^{-1/5} C_n^{6/5} L^{8/5}) \tag{2.12}$$

The wavelength dependency on beam spreading is not strong. The spot size can often be observed to be twice that of the diffraction-limited beam diameter. Many FSO systems incur approximately 1 m of beam spread per kilometer of distance.

2.2.4   Ambient Light

Natural light sources, like the sun and moon light, have spectral lines in the visible and infrared region, and therefore, induce noise in the FSO links. The modeling of noise in FSO system will be discussed in later chapters. The solar radiations spectrum extends from 300 nm to more than 1500 nm with varying intensities. The peak intensity is located around 480 nm which progressively decrease as the wavelength increases, with the spectrum shown in Fig. 2.7 [8].

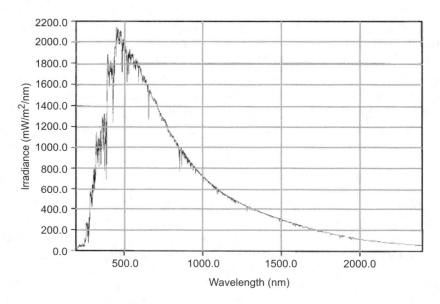

Figure 2.7   Absolute Solar Spectral Irradiance.[8]

Besides the above atmospheric factors of attenuation, turbulence and sunlight effects, other causes of signal attenuation in FSO terrestrial links are:

➤ *Physical obstruction:* Optical beam can be totally disrupted by any physical obstruction in the line of sight path. Flying birds or any object in the line of sight path can temporarily block a single-beam in FSO system, but this tends to cause only short interruptions, and transmissions are easily and automatically resumed.

➤ *Window glass-pane loss:* If the FSO transmitter and receiver are installed behind a glass window, then the loss of the window glass-pane will be there, which also depends on wavelength. Window loss includes two major loss components; one due to reflection from the air-glass interface and another due to the glass absorption. A tinted window gives larger loss and is not recommended for FSO systems.

➤ *Building Sway/seismic activity:* FSO beams are quite narrow, misalignment due to building sway can interrupt the communication link. For an FSO link alignment, it is necessary to ensure that the transmit beam divergence angle not only matches up with the field of view of the receive telescope, but also takes into account the misalignment due to building sway and seismic activities in the nature.

2.3 WIRELESS OPTICAL SPACE LINKS

For space links, near earth or outer space, one has to consider the atmosphere around the earth. The concentric layers around the earth are broadly divided in two regions: The *Homosphere* and the *Heterosphere.* The heterosphere lies above the homosphere. The homosphere, which is from 0 km to around 90 km, has three layers differentiated by the temperature gradient with respect to altitude:

➤ Troposphere

➤ Stratosphere and

➤ Mesosphere

In the case of wireless optical space link performance, we are mainly interested in the troposphere region of the homosphere, as this is where almost all the weather phenomena described above occur and cause attenuation in the light transmission. Rest of layers have only the free space loss. Therefore, the cause of beam power attenuation in these links is once again; the *beam divergence* and the *atmospheric losses.* The atmospheric loss components, as discussed above, are of both, fixed and of variable nature. Components with fixed attenuation are proportional to the density of N_2, O_2, Ar and CO_2. They have almost quasi uniform distribution ranging from 15–20 km. The other components with variable density are less and their concentration depends on geographic location, on environment and weather condition. The main constituents of this category are the water vapor and aerosols and their presence is negligible beyond 20 km.

The Stratosphere and Mesosphere have no atmosphere and do not contribute to any loss, but do cause the divergence of the beam. Also, noise due to solar radiation is contributed by these layers.

2.4 INDOOR PROPAGATION MEDIA

Indoor atmosphere is free of environmental degradation, such as rain, fog, particulate matter, clouds, etc. Therefore, indoor optical wireless systems encounter only the free space loss of beam divergence, noise due to ambient light and signal fading due to multipath reception of signals.

Free Space Loss   It is that part of the transmitted power, which is lost or not captured by the receiver's aperture (Fig.2.2) due to beam propagation loss and beam divergence, respectively. A typical figure for a point-to-point system, that operates with a slightly diverging beam, would be 20 dB, whereas an indoor system using a wide-angle beam in case of diffused links could have a free space loss of 40 dB or more [9].

Noise in Indoor System   The dominant source of noise in indoor optical wireless systems is ambient light, which is due to combination of fluorescent light, sunlight, and incandescent light. Also, typical intensity levels of the ambient light collected at the photo detector are usually much higher than data signal intensity levels. Besides the ambient light producing shot noise, in the indoor systems we also have additional noise due to lighting in the rooms. The fluorescent and incandescent lamps produce interference due to periodic variations of the light intensity. These variations can occur at a frequency double the power line frequency, and at the switching frequency of electronic ballasts of the fluorescent lamps. Figure 2.8 gives the spectrum of the principal sources of noise, i.e., due to fluorescent lamps, incandescent lamps and daylight [10]. In general, for low and moderate data rates the ambient noise is the main factor degrading the performance of wireless indoor systems.

Multipath Fading   In the non-LOS diffused systems the laser beam reaches the receiver after multiple reflections from walls and other obstacles. This causes the light to reach in different phases and hence cause interference and fading in the received signal [10] leading to inter symbol interference at high data rates.

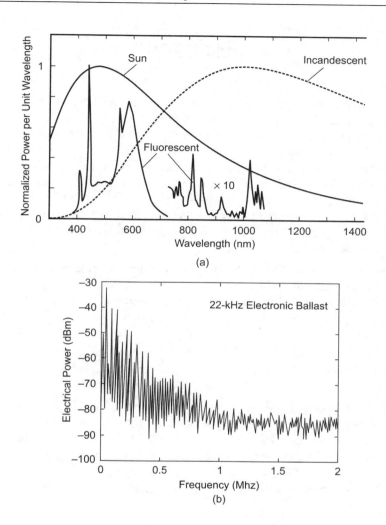

(a)

(b)

Figure 2.8 (a) Optical Power Spectra of Common Ambient Infrared Sources. Spectra have been Scaled to have the Same Maximum Value. (b) Detected Electrical Power Spectrum of Infrared Emission from a Fluorescent Lamp Driven by 22-kHz Electronic Ballast [11].

Summary

Outdoor free Space optical communication systems are affected by variety of atmospheric effects. There is attenuation due to absorption of laser beam by the constituent molecules and aerosol particles in the atmosphere. Scattering by these particles reduces the energy of the beam directed towards the receiver and hence causes attenuation. Fog contributes maximum attenuation, both due to scattering as well as absorption which is followed by snow, rain and dust. Atmosphere has low-loss transmission windows with attenuation as low as 0.2 dB/Km within the 0.8 to 1.6 micron wavelength region in clear air due to the spectral response of the gaseous molecules. Fortunately this is also the low attenuation window of optical fiber.

Besides the attenuation due to the molecules and aerosols, optical turbulence results from small temperature variations giving rise to power losses from beam spreading beyond that due to diffraction alone and spatial and temporal fluctuations of the laser beam known as scintillation. The effect of turbulence is random in nature.

The indoor wireless systems fortunately do not have atmospheric attenuation, but they are affected by the ambient light and fluorescent light noise. Non-LOS systems are also affected by the multipath effect.

Air Force Research Laboratories (AFRL), Space Vehicle Directorate and Spectral Science, Inc have provided software with the atmospheric Transfer Code System and algorithms titled MODTRAN (MODerate Spectral Resolution Atmospheric TRANsmittance). The software provides the transmission radiance at the wavelengths of interest. It can be used to predict the laser signal radiation through the atmosphere in most weather conditions along a terrestrial path [12].

References

1. Friedman E., Miller John L., *Photonics Rules of Thumb*, McGraw-Hill Professional, 2003.
2. Tyson R., *Principles of Adaptive Optics*, Academic Press, San Diego, CA, 1991.
3. Weichel H., *Laser Beam Propagation in the Atmosphere*, SPIE, Bellingham WA, 1990.
4. Pratt W.K., *Laser Communication Systems*, J. Wiley & Sons, New York, 1969.

5. Kim, Issac I., Arthur Bruce and Korevaar E., 'Comparison of laser beam propagation at 785 nm and 1550 nm in fog and haze for optical wireless communications,' Optical wireless communication conference, 2001, SPIE, vol. 4214, pp. 26-37.

6. Wallace J.M. and Hobbs P.V., *Atmospheric Science: An Introductory Survey*, Academic Press, Orlando, 1977.

7. Al-habasch A., Fischer, K.W., Cornish C.S., et.al., 'Comparision between experimental and theoretical probability of fade for free space optical communications', Optical wireless communication V, Proceedings of SPIE, 4873.79–89, 2002.

8. Thuillieri G., Herse M., Labs D., Foujols T., Peetermans W., Gillotay D., Simon P.C. and Mandel H. 'The solar spectral irradiance from 200 to 2400 nm as measured by the solspec spectrometer from the atlas and eureka missions 2' Solar Physics 214: 1–22, 2003 Kluwer Academic Publishers, Netherlands.

9. Raminez-Inignez R., Greene R.J., *Indoor optical wireless communication*, IEE, Sovoy Place, London WC2ROBl, UK, 1999.

10. Kahn J. M., and Barry J.R. 'Wireless Infrared Communications', *Proceedings of the IEEE*, Vol. 85, No. 2, Feb. 1997, pp. 265-298.

11. Narasimhan R., Audeh M.D., and Kahn J.M., 'Effect of electronic-ballast fluorescent lighting on wireless infrared links,' IEE Proc.-Optoelectron., Dec. 1996.

12. Smith F.G., Accetta J.S., Shumaker D.L., *The Infrared and Electro-Optical Systems Handbook. Atmospheric Propagation of Radiation*, Vol. 2., SPIE Press, 1993.

Further Reading

1. L.C. Andrews and R. L. Phillips, *Laser Beam Propagation through Random Media*, 2nd Ed. SPIE Press, 2005.

2. L.C. Andrews, R. L. Phillips, and C. Y. Hopen, *Laser Beam Scintillation with Applications* , SPIE Press, 2001.

3. L.C. Andrews, et al., *Beam Wander Effects on the Scintillation Index of a Focused Beam*, SPIE Press, 2005.

4. A.C. Boucouvalas, 'Indoor ambient light noise and its effect on wireless optical links,' IEE Proceedings-Optoelectronics, vol. 143, pp. 334-338, 1996.

5. Gebhart, M. Leitgeb, E. Bregenzer, J., 'Atmospheric effects on optical wireless links', ConTEL2003. Proc. of the 7th International Conference on Communication. 11-13, June 2003, Vol:2, 395-401.

6. Jingzhi Wu,; Yangjun Li, 'Atmospheric effects on wireless optical communications' ICMIT 2005: Information Systems and Signal Processing, Proc. Vol. 6041, 20 February 2006.

7. Shlomi Arnon, 'Effects of atmospheric turbulence and building sway on optical wireless-communication systems', *Optics Letters*, Vol. 28, Issue 2, pp. 129-131.

8. Ghassemlooy, Z. and Popoola, W. O. 'Terrestrial Free-Space Optical Communications', in *Mobile and Wireless Communications Network Layer and Circuit Level Design,* Edited by: Salma Ait Fares and Fumiyuki Adachi, Pub: InTech, Jan. 2010.

Chapter **3**

Channel Modeling

The signal response of the optical beam is affected by the channel characteristics discussed in the last chapter. While designing any communication system, few questions are significant, i.e., the range of power levels, the bit rates necessary for a given link length, the types of modulation and detection techniques to be used for good performance, etc. To a certain extent, these questions can be answered by collecting some experimental data, but that cannot be generalized to all situations. It certainly helps to know theoretically what signal response to expect from the channel. The endeavor in this chapter is to obtain the response of the atmospheric channel in terms of theoretical channel modeling. Similar to RF wireless channel there are probabilistic channel models available describing the physical mechanisms of the atmosphere at optical wavelengths. Even if these are over-simplified models, but we need them to compare different system approaches to get a sense of what types of approaches are worth pursuing, etc. There are many probabilistic models for optical wireless channels available in literature, and they have been highly useful for providing insight into these systems.

In this chapter, we first explain in Section 3.1 the types of signal variation we get when the laser beam propagates through the atmospheric channel. Then starting with the constraints of optical wireless system, the channel impulse response is derived for a base-band linear time varying system in Section 3.2. In Section 3.3, we do the mathematical modeling of the optical channel and then construct stochastic models of the channel with the assumption that different channel behavior appear with different probabilities and change over time. The impulse response is then derived for this random statistical atmospheric model. The channel models presented in this chapter are for both, indoor (LOS and diffuse channel), and for the outdoor systems. Section 3.4 derives the system optical transfer function and Section 3.5 gives the channel modeling for indoor systems. The noise sources present in FSO are modeled in Section 3.6.

3.1 LARGE AND SMALL-SCALE SIGNAL VARIATION

The physical phenomenon discussed in the last chapter causes variation of the signal strength over time and frequency with distance. These variations define the characteristic of the optical wireless channel. The time variations can be broadly categorized as: *Large signal variation* and *small-scale signal variation.* The large signal variation is due to the atmospheric attenuation of the signal which is caused by fog, rain, etc., The small-scale signal variation or *fading* is due to the turbulence in the atmosphere, or the constructive and destructive interference of the multiple signal paths between the transmitter and receiver. Large signal variation is typically frequency independent once the low-loss window is selected. This was discussed in the previous chapter. It is taken into account for link power budget design and, therefore, the link length and the transmitted power level get decided once the power budget consideration are taken into account. On the other hand, the small-scale fading is time- and wavelength-dependent, and occurs at the spatial scale of the order of the optical carrier wavelength. The small-scale fading, which is due to multi-path or turbulence, is random in nature and is very important for the design of reliable and efficient communication systems. We will discuss the small-scale fading in this chapter.

As in the other cases of unconfined wireless transmission systems, multi-path propagation effects are important in wireless optical networks as well. Multi-path is a common phenomenon, which occurs in a transmission link when the transmitted signal follows different paths on its way to the receiver either due to its reflection by walls, ceilings, etc., or due to scattering in the atmosphere. The reception of signals via different paths by the receiver can cause signal fading. Some of signals interfere destructively and thus the received signal power effectively decreases. This can be observed in both indoor and outdoor optical wireless systems.

Unlike radio systems, multi-path fading to a certain extent is mitigated in wireless optical transmission. This fortunately happens due to *antenna averaging* or the inherent *spatial diversity* in the receiver. The aperture of the light detector, which is the receiving antenna in the wireless optical system, has an active radiation collection area of the order of 10^4–10^5 times of the wavelength of the infrared light. This large size of the detector with respect to the wavelength provides the antenna averaging in the receiver which mitigates the effect of multi-path fading [1]. The wireless optical links though still suffer from *temporal dispersion* of the received signal due to multi-path propagation. This causes inter-symbol-interference (ISI). In the indoor systems the diffused links are more prone to multi-path effects than the directed beam systems. This is because of their larger beam widths of the transmitter and the larger FOV of their detector, resulting in collecting reflected light from more number of reflectors and collecting at the detector. In the terrestrial links, as the rate of transmission increases above 10 Mbps, the ISI caused by the multi-path dispersion becomes a major degrading factor.

3.2 INPUT/OUTPUT MODEL OF THE OPTICAL INTENSITY WIRELESS CHANNEL

The optical signal is an electromagnetic signal. Therefore, the E- H- field of the signal decays inversely with distance, and the optical power as inverse square of the distance. In practice, because of the atmosphere and several obstacles between the transmitter and the receiver, which might absorb some power while scattering the rest, one expects the power to decay considerably faster. Indeed, at large distances the power decays exponentially with distance. As in practice we are looking for the order of decay of power with distance. Therefore, instead of electromagnetic analysis, which may be computationally very rigorous for the complicated problem of real life link and the surrounding atmosphere, we can consider an alternative approach. We look for a model of the physical environment with the fewest number of parameters to statistically model the channel.

3.2.1   Constraints of Optical Intensity Channel

In order to derive the input/output model of the channel in this section we first need to know the constraints of the optical channel. Figure 3.1 gives the schematic of an optical wireless system, which is considered to be intensity modulated direct detection system (IM-DD) as commonly employed in FSO terrestrial links. In the transmitter the LED or LD is biased above threshold for linear conversion between the input drive current, $x(t)$, and the output optical power, $P_t(t)$. The light detector is a square law device which integrates the square of the amplitude of the electromagnetic radiation impinging on it or there is linear conversion of optical power. The electro-optical conversion at the transmitter can be modeled as $P_t(t) = mx(t)$, where m is the optical gain of the device with units of W/A. On the receiving side also, the linear opto-electronic conversion of the received optical power signal $P_r(t)$ to the detector electrical photocurrent $y(t)$ is modeled as $y(t) = \Re P_r(t)$, where, $\Re$ is the detector responsivity in units of A/W. The received signal at the receiver is written as $P_r(t) = \alpha_0(\lambda,t)P_t(t)$ where α_0 depends on the link length, characteristics of the channel and optics used. We use a model of incoherent propagation and as a result the characterization of light propagation is based upon a power-intensity distribution and not upon an electromagnetic field-amplitude distribution. From the above relations, the channel output P_r can be written in terms of the channel input as:

$$P_r(t) = y(t)/\Re = \alpha_0(\lambda,t).m.x(t)$$

$$\text{or} \quad y(t) = \alpha_0(\lambda,t).m.\Re x(t) \tag{3.1}$$

Eq. (3.1) expresses the channel in terms of the base-band electrical quantities of input drive current of the LD and the output detector current. The values of m and $\Re$ are constant and we can assume their product to be unity for the purpose of solution of (3.1).

The optical system differ from conventional electrical system in few respects. One, the channel input in the optical case represents instantaneous optical power, i.e., the channel input is nonnegative, or

$$P_t(t) \geq 0 \tag{3.2}$$

This leads to the non-negativity constraint on $x(t)$ as well. The second constraint is that the average transmitted optical power P_t is limited to P, which is the average power fixed due to eye-safety consideration,

$$P_t = \lim_{T \to \infty} \frac{1}{2T} \int_{t=0}^{\infty} P_t(t)dt \le P \qquad (3.3)$$

Equivalently, in the optical case, the average amplitude of $x(t)$ is upper bounded by P rather than the time-average of $x^2(t)$, as in the RF case when $x(t)$ represents amplitude of the signal.

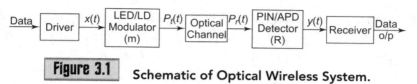

Figure 3.1 Schematic of Optical Wireless System.

The optical channel with atmospheric turbulence and multi-path effects can be modeled as a linear time varying system. In the following section, therefore, we first obtain the impulse response of the linear time varying system model and subsequently obtain the impulse response for the outdoor and indoor channels.

3.2.2 Linear Time-Varying System Model of the Channel

The received signal at any point in the channel is in response to a weighted sum of all the transmitted waveforms. The linear relationship between input and the received signal is a consequence of the fact that the received signal consists of many electromagnetic modes of the optical signal. Let us first consider, in general, the response of the system to the modulating signal on optical carrier. The received signal is a power signal with $\alpha_i(\lambda,t)$ as the overall attenuation of the path i from the transmitter to the receiver at λ optical wavelength and $\tau_i(\lambda,t)$ their respective delays at time t. The overall attenuation is simply the product of the attenuation factors due to the distance from the transmitting antenna to the receive antenna, and the beam directionality of the transmitter and the capture area of the detector at the receiver. Though we have described above the channel effects at a particular wavelength, but if we assume that $\alpha_i(\lambda,t)$'s and $\tau_i(\lambda,t)$ do not depend on the wavelength but are rather constant over the waveband once the transmission window is selected, then we can use the principle of superposition to generalize the input-output relation. In fact, the attenuations and the propagation delays in practice are usually slowly varying functions of wavelength hence the assumption is valid.

In any case, we are primarily interested in transmitting over bands that are narrow relative to the selected optical carrier frequency, and over such ranges we can omit this wavelength dependence. Hence, the input-output relation of (3.1) to an arbitrary input drive current $x(t)$ with nonzero bandwidth is:

$$y(t) = \sum \alpha_i(t)x(t - \tau_i(t)) \tag{3.4}$$

Since the channel in equation (3.4) is linear, it can be described by the channel response $h(t)$, at time t to an impulse transmitted at time $(t - \tau)$. The input-output relationship in terms of $h(\tau, t)$, is given by:

$$y(t) = \int_{-\infty}^{\infty} h(t, \tau)x(t - \tau)d\tau \tag{3.5}$$

Comparing equations (3.4) and (3.5), we see that the impulse response of the base-band equivalent optical channel is:

$$h(\tau, t) = \sum_i \alpha_i(t)\delta(t - \tau_i(t)) \tag{3.6}$$

Equation (3.5) very elegantly gives the relationship of the input/output quantities of the channel between the transmitter and the receiver without going through the ordeal of solving the Maxwell's equations. This is simply represented as the impulse response of a linear time-varying channel filter, which takes into account the effect of multi-path in the atmosphere due to turbulence and other obstacles.

In the special case when the transmitter, receiver and the environment are all stationary, the attenuations $\alpha_i(t)$'s and propagation delays $\tau_i(t)$'s do not depend on time t, and we have the linear time-invariant channel with an impulse response.

$$h(\tau) = \sum_i \alpha_i\delta(\tau - \tau_i) \tag{3.7}$$

Next, from the impulse response, $h(\tau)$, we can define the frequency response of the channel by its Fourier Transform [2]:

$$H(f) = \int_{-\infty}^{\infty} h(\tau)\exp(-2\pi f\tau)d\tau = \sum_i \alpha_i \exp(-j2\pi f\tau_i) \tag{3.8}$$

It is usually appropriate to model the channel $h(t) \leftrightarrow H(f)$ as fixed, since it normally changes only when the transmitter, receiver in the terrestrial link or objects in the room for the indoor system are moved by tens of centimeters.

3.2.3 Channel Transfer Function in Optical Domain

In terms of electrical quantities, i.e., the diode drive current and the photodetector current $x(t)$ and $y(t)$, respectively, the channel transfer function was defined as the Fourier transform of the impulse response. At the optical intensity level, the Optical Transfer Function (OTF) is defined in terms of spatial frequency, which in one-dimension is defined as lines per unit width. The channel OTF is the normalized FT of the average light intensity distribution in the image plane as caused by a point object at the object plane. The OTF is defined as follows [3]:

$$\text{OTF} = \xi(f_x, f_y) = \frac{\displaystyle\iint_{-\infty}^{\infty} s(x', y') \exp[-j(f_x x' + f_y y')] dx' dy'}{\displaystyle\iint_{-\infty}^{\infty} s(x', y') dx', dy'} = \frac{S(f_x, f_y)}{S(0,0)} \quad (3.9)$$

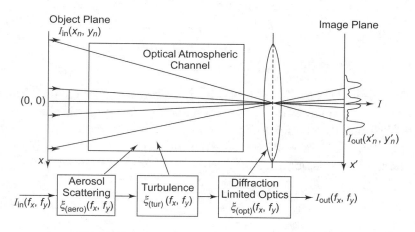

Figure 3.2   Optical Transfer Function Model.

where, $S(f_x, f_y)$ is the two dimensional Fourier Transfer Function of the Point Spreading Function (PSF) $s(x', y')$, $S(0,0)$ is the maximum value of the OTF which occurs at zero spatial frequency, and (f_x, f_y) are the spatial frequencies in the x- and y- direction. The OTF in (3.9) can be separated into two components: magnitude and phase and is expressed as:

$$\text{OTF} = \text{MTF} \exp (jPTF)$$

The magnitude component, known as the Modulation Transfer Function (MTF), is responsible for the size of the received signal in terms of spatial coordinates. The phase component, known as Phase Transfer

Function (PTF) determines pulse position and orientation rather than size. It is less important than the MTF, but often cannot be neglected.

Similar to electrical transfer function, the OTF relates the output and the input optical intensity as:

$$I_{out}(f_x, f_y) = \xi(f_x, f_y)I_{in}(f_x, f_y) \tag{3.10}$$

where, I_{in} (f_x, f_y) and $I_{out}(f_x, f_y)$ are intensities of the input and output signals of the channel in the spatial frequency domain. If we assume that the emitting element in the object plane is a spatial δ function, positioned at (x_1, y_1) locations in the x and y coordinates [4] as shown in Fig. 3.2, it is expressed as:

$$I_{in}(x, y) = I_0 \delta(x - x_1, y - y_1) \tag{3.11}$$

In the spatial frequency domain, (3.11) can be rewritten as:

$$I_{in}(f_x, f_y) = I_0 \exp(-j2\pi f_r r_1)$$

$$f_r = \sqrt{(f_x^2 + f_y^2)}, \ r_1 = \sqrt{(x_1^2 + y_1^2)} \tag{3.12}$$

The channel OTF $\xi(f_x, f_y)$, depends on the atmospheric channel and the optics at the transmitter and the receiver. In order to obtain the intensity of the beam in the spatial domain in the image plane, from (3.9) we obtain the point spread function PSF, which indicates the channel's transference of intensity. This is obtained by the inverse FT of the channel's OTF given by;

$$\varsigma(x_1', y_1') = F^{-1}\{\xi(f_x, f_y)\} \tag{3.13}$$

Then the output signal at the photodetector can be expressed as:

$$I_{out}(x_1', y_1') = \varsigma(x_1', y_1') * I_{in}(x_1, y_1), \text{ where } x_1' = \frac{f}{L}x_1$$

where x_1', y_1' are the coordinates in the image plane at the receiver, L is the link length and 'f' is the focal length of the lens at the receiver.

3.2.4 Additive White Noise

In the calculation of channel response, we have yet not included the noise added in the signal by the channel. As a last step we now include the noise $w(t)$. We make the assumption that $w(t)$ is *zero-mean, additive, white Gaussian noise* (AWGN) with power spectral density $N_0/2$. Including the noise, the model (3.4) is now modified to be:

$$y(t) = \sum_i \alpha_i(t) x(t) + w(t) \tag{3.14}$$

The assumption of AWGN essentially means that we are assuming that the primary source of the noise in the channel is either at the receiver or is in the radiation impinging on the receiver that is independent of the paths over which the signal is being received. This is a reasonably good assumption for both outdoor and indoor channels.

3.3 STATISTICAL CHANNEL MODELING OF FSO

Multi-path and fading due to atmospheric turbulence is a random phenomenon. The base-band impulse response of this channel, therefore, will be random, and hence a statistical characterization will be required. Before we start the details of the stochastic models we familiarize ourselves with few of the terms which are used in this context with reference to time and frequency variation of the channel.

3.3.1 Parameters of the Wireless Optical Channel

The parameters of the optical wireless channel, which determine its time and variation of frequency, are the *Coherence Time* and *Coherence bandwidth,* respectively.

Coherence time is an important channel characteristic which determines the time-scale variation of the channel; how fast the channel change with time. For the baseband channel it is determined by variation of $h(\tau,t)$, which is a random process and depends on how fast does it vary as a function of time t. We know that:

$$h(\tau,t) = \sum_i \alpha_i(t)\delta(\tau - \tau_i(t)) \tag{3.15}$$

where, $\alpha_i(t) = |\alpha_i| \exp(-j2\pi f \tau_i(t))$; with $f = c/\lambda$

The coherence time T_c, of a wireless channel is defined as the interval over which $h(\tau,\ t)$ changes significantly as a function of time. Coherence time can be approximately obtained as:

$$T_c \approx \frac{4}{D_s} \tag{3.16}$$

where, D_s measures the spectral broadening caused by the time rate of change of the channel. It is defined as the time-scale, which is inversely proportional to the largest difference between the phase-shifts given by:

$$D_s = \max_{i,j} f_c \left| \tau_i(t) - \tau_j(t) \right| \tag{3.17}$$

where the maximum is taken over all the paths that contribute significantly. It is observed that there is no significant change in amplitude of α in equation (3.15) over periods of seconds or more, but the phase of the ith path changes significantly. The phase change of 2π or more radians can occur with relatively small changes in the medium characteristics or path change as the optical frequencies are in terahertz range. Essentially, it means that it takes much longer for a path or the channel to change in magnitude than for its phase to change significantly. These phase changes are significant over delay changes of the order of $1/(4D_s)$.

Often in the wireless communication, channels are categorized as *fast fading* and *slow fading*. A channel is said to be fast fading if the coherence time T_c is much shorter than the delay requirement of the application, i.e., one can transmit the coded symbols over multiple fades of the channel. In the case of slow fading, T_c is longer than the delay requirement of the application. Thus, whether a channel fading is fast or slow depends not only on the channel characteristic but also on the application. For example, for voice, the delay requirement is short, typically less than 100 ms, while on the other hand for data applications more relaxed delay requirement are permissible.

Coherence bandwidth characterizes the frequency-scale variation of the channel: Wireless channels change, both in time and frequency. While the time coherence shows how quickly the channel changes in time, the frequency coherence, on the other hand, shows how quickly it changes with frequency. We can see this in terms of the frequency response of the channel. The frequency response at time t, as expressed in equation (3.8) is given as:

$$H(f) = \sum_i \alpha_i(t) \exp(-j2\pi f \tau_i(t))$$

From the above, we observe that though the phase changes linearly with frequency due to a particular path, but there is a differential phase of $2\pi f(\tau_i(t) - \tau_k(t))$ for multiple paths. It is this differential phase, which causes selective fading with frequency. The coherence bandwidth, W_c, is defined as:

$$W_c = \frac{1}{2T_d} \tag{3.18}$$

where, T_d is the multi-path *delay spread*, defined as the difference in propagation time between the longest and shortest path, counting only the paths with significant energy. Thus:

$$T_d = \max_{i,j} \left| \tau_i(t) - \tau_j(t) \right| \tag{3.19}$$

The path delays are of the order of picoseconds at optical frequency. As links become smaller, T_d also shrinks. When the bandwidth of the input is considerably less than W_c, the channel is usually referred to as *flat fading*. In this case, the delay spread T_d is much less than the symbol time T_s. When the bandwidth is much larger than Wc, the channel is said to be *frequency-selective*. As in the case of time variation, flat or frequency-selective fading is not the property of the channel alone, but it depends on the relationship between the bandwidth of the signal and the coherence bandwidth Wc.

Once T_c and Wc of the channel are determined, the need for detail study of the propagation over multiple paths and the complicated types of reflection/scattering mechanisms is not required to obtain the characteristics of the channel. For the design of the system all, we really need is the aggregate values of gross physical mechanisms, such as spectral spread and multi-path delay spread.

3.3.2 Modeling of Atmospheric Turbulence

The fading or the random fluctuation in the optical wireless channel response $h(t)$ is due to the multi-path and turbulence in the medium. Therefore, stochastic modeling of the atmospheric turbulence is the next task to obtain the channel model.

The atmospheric turbulence is described by Kolmogorov theory [5-7] according to which turbulence is due to random motion of large regions or cells of nearly the same refractive index. These start out quite large, but due to atmospheric mixing, the energy of large cells/eddies is redistributed without loss to eddies of decreasing size. The size of turbulence eddies normally ranges from a few millimeters to a few meters, denoted as the inner scale l_0 and the outer scale L_0, respectively. As the light beam travels through these eddies, the random variation in the refractive index perturbs the velocity of the wavefront, giving rise to randomness in the received signal at the receiver. The resulting interference causes fading of the received signal power. With reference to atmospheric turbulence, the atmosphere is characterized one, by the measure of the eddy size and two, by the atmospheric structure constant C_n^2, also called the wave number spectrum structure parameter at a given time and location.

The refractive index of the atmosphere can be given by [8]:

$$n(\overline{r},t) = n_0 + n_1(\overline{r},t) \tag{3.20}$$

where, n_0 is the average index and n_1 is the fluctuation component induced by the atmospheric turbulence. The variation of n_1 with space and time can be expressed by its correlation function as:

$$\Gamma_{n1}(\overline{r}_1,t_1;\overline{r}_2,t_2) = E[n_1(\overline{r}_1,t_1).n_1(\overline{r}_2,t_2)] \tag{3.21}$$

For the time-invariant system the correlation function reduces to $\Gamma_{n1}(\overline{r}_1;\overline{r}_2)$, which describes the spatial coherence of the refractive index. Many models have been proposed to study the spatial coherence of the refractive index. These are exponential, Gaussian or other solvable function forms.

The spatial Fourier transform of $\Gamma_{n1}(\overline{r}_1;\overline{r}_2)$ is denoted as $\varphi_n(\tilde{k})$. It was proposed by Kolmogorov by a widely used model with a good accuracy as:

$$\varphi_n(\tilde{k}) = 0.333 C_n^2 k^{-11/3} \tag{3.22}$$

where the wave-number spectrum structure parameter C_n^2 is altitude dependent. For horizontal path up to the heights of few kilometers the value of C_n^2 is taken to be constant. For downlink or uplink in optical space links the C_n^2 profile model changes with altitude. Hufnagel and Stanley [9] gave a simple model for C_n^2 with altitude variation as:

$$C_n^2(h) = K_0 h^{-1/3} \exp(-h/h_0) \tag{3.23}$$

where, K_0 is the parameter describing the strength of turbulence and h_0 is the effective height of the turbulent atmosphere. For atmospheric channels near the ground ($h < 18.5$m), C_n^2 can vary from $10^{-13} m^{-2/3}$ to $10^{-17} m^{-2/3}$ for strong to weak turbulence, respectively. On an average the value of C_n^2 is $10^{-15} m^{-2/3}$. The C_n^2 parameter with wind velocity included can be expressed the Hufnagel Valley formula as [10]:

$$C_n^2(h) = 0.00594 \left(\frac{V}{27}\right)^2 (10^{-5}h)^{10} \exp\left(\frac{-h}{1000}\right)$$
$$+ 2.7 \times 10^{-16} \exp\left(\frac{-h}{1500}\right) + A \exp\left(\frac{-h}{100}\right) \tag{3.24}$$

where, h is the altitude in meters, V is the rms wind speed in meters per second and A is a nominal value of $C_n^2(0)$ at the ground in $m^{-2/3}$.

3.3.3 Spatial Coherence of Optical Signals

The variation of the refractive index due to turbulence affects the spatial and temporal coherence of optical signals, which in turn causes scintillation in the light intensity defined by the scintillation index as:

$$\sigma_I^2 = \frac{\langle I^2 \rangle - \langle I \rangle^2}{\langle I \rangle^2} \tag{3.25}$$

where, $\langle I \rangle$ is the ensemble average of irradiance of the optical wave. Assuming the process to be ergodic it is equal to the long-term average.

The Mutual Coherence Function (MCF) [11] is used to describe the spatial coherence of optical waves. For the time-invariant channel, the spatial MCF is expressed as:

$$\Gamma_u(r_1; r_2) = E[u(r_1).u^*(r_2)] \tag{3.26}$$

where $u(r)$ is the complex optical field. Next, we need to relate the refractive index model to an EM model; Rytov assumed the field at any point in the medium as the product of free space field and the field perturbation given by the stochastic complex amplitude transmittance. Accordingly, by the Rytov method, the optical field $u(\tilde{r})$ is given by:

$$u(\tilde{r}) = A(\tilde{r}).\exp(j\Phi(\tilde{r})) = u_0(\tilde{r}).\exp(\Phi_1) \tag{3.27}$$

where, $u_0(\tilde{r})$ is the complex field without air turbulence and is expressed as:

$$u_0(\tilde{r}) = A_0(\tilde{r}).\exp\left[j\Phi_0(\tilde{r})\right] \tag{3.28}$$

The perturbation factor in the exponent of equation (3.27) is given by:

$$\Phi_1 = \log\left[\frac{A(\tilde{r})}{A_0(\tilde{r})}\right] + j\left[\Phi(\tilde{r}) - \Phi_0(\tilde{r})\right] = X + jS \tag{3.29}$$

where, X is the log-amplitude fluctuation, and S is the phase fluctuation. For long propagation distances through turbulence, it is valid to assume the log-amplitude X, and the phase fluctuation S, as homogeneous, isotropic and independent Gaussian random variables.

Next, to obtain the spatial variation of the turbulence induced fading of log-amplitude X, the covariance function of X is obtained as [11]:

$$B_X(r_1, r_2) = E[X(r_1).X(r_2)] - E[X(r_1)].E[X(r_2)] \tag{3.30}$$

With the assumption of weak turbulence and using the Rytov method, the normalized covariance function of X for two positions in a receiving plane perpendicular to the distribution of propagation can be expressed by as:

$$b_X(d_{12}) = \frac{B_X(r_1, r_2)}{B_X(r_1, r_1)}$$

(3.31)

where, d_{12} is the distance between the two positions, r_1 and r_2. It is expressed by a useful parameter, the *correlation length d_0*. The correlation length of intensity fluctuation is defined as the distance for which, $b_X(d_0) = e^{-2}$. This describes the intensity fluctuations of turbulence induced fading in the plane of the receiver aperture. When the propagation distance L satisfies the condition: $l_0 < sqrt(\lambda L) < L_0$, d_0 can be approximated by [7]:

$$d_0 = \sqrt{\lambda L}$$

(3.32)

where, λ is the wavelength and l_0 and L_0 are the inner and outer length scales defined in Section 3.3.2. In most free space optical communication systems with visible or infrared lasers and with propagation distance of a few hundred meters to a few kilometers, equation (3.32) is valid. When the receiver aperture is made larger than the correlation length, then turbulence induced fading can be reduced substantially by *antenna averaging*.

3.3.4 Channel Impulse Response/Probability Distribution of Turbulence-Induced Intensity Fading

Electromagnetic waves at optical wavelength, when propagate through the atmospheric turbulence, suffer temporal and spatial fluctuations in the form of scintillation as discussed in the last sections. The reliability of the designed optical communication link will depend on the accurate estimate of the random fluctuation in the signal or in other words on the probability density function (pdf) of this random irradiance signal, which is the channel impulse response $h(t)$. There are many pdf models proposed for both weak and strong turbulence. From practical point of view, it is desirable to have a tractable pdf model for the irradiance fluctuations so that we can predict the performance of the link with an acceptable accuracy. The pdf's proposed for these channels for different degree of turbulence are:

➢ Log-normal

➢ Negative exponential[11]

➢ I-K distribution[12]

➢ Log-normal Rician[13]

➢ Gamma-Gamma[14]

Log-normal distribution is the most commonly used model for the probability density function of the irradiance for weak turbulence condition. But log-normal pdf underestimates the behavior in the tails as compared with the measured data. In high turbulence conditions, there are multiple scattering effects, which become important. These cannot be accounted accurately with log-normal pdf, which significantly affects the accuracy of the estimated performance at high turbulence. The accuracy of detection and fade probability primarily depends on the tails of the pdf and, therefore, underestimating in this region will give inaccurate performance results [12]. The negative exponential pdf, on the other hand, models high turbulence effects well. The rest three distributions have the irradiance modeled as the result of two multiplicative random processes. Gamma-gamma is a two-parameter distribution based on a double stochastic theory and can model both, atmospheric weak and strong turbulence conditions. The smaller intensity variation and the larger fluctuation are governed by independent gamma distribution. The large irradiation fluctuation modulates the smaller intensity variation. In the following sections we will be discussing in detail the pdf models, which are more commonly used in practice for modeling the optical channel.

3.3.4.1 *Log-normal Fading and Negative Exponential Models*

For propagation distances less than few kilometers, variations of the log-amplitude X, are typically much smaller than variations of the phase in equation (3.29). Over longer propagation distances, where turbulence becomes more severe, the variation of the log-amplitude X can become comparable to that of the phase S. Based on the atmospheric turbulence model defined in the earlier section and assuming weak turbulence, the amplitude variation of the beam irradiance are given by the *log-normal distribution*.

When the propagation of light is through a large number of elements of the atmosphere, each causing an independent, identically distributed (i.i.d) phase delay and scattering, then by Central Limit Theorem the pdf of the log-amplitude X is *Gaussian* and is given as:

$$f_X(X) = \frac{1}{\left(\sqrt{2\pi\sigma_X^2}\right)}\exp\left\{-\frac{(X-\langle X\rangle)^2}{2\sigma_X^2}\right\} \quad \text{with } X > 0 \qquad (3.33)$$

The intensity of light is related to log-amplitude, X by:

$$I = I_0\exp(2X - 2\langle X\rangle) \qquad (3.34)$$

where, $\langle X\rangle$ is the ensemble average. From equations (3.33) and (3.34), the marginal distribution (i.e., fading at a single point in space at a single instant in time) of light intensity fading induced by turbulence is *log-normal*:

$$f_I(I) = \frac{1}{2I}\frac{1}{\left(2\pi\sigma_X^2\right)^{1/2}}\exp\left\{-\frac{\left[\ln(I)-[\ln(I_0)]\right]^2}{8\sigma_X^2}\right\} \qquad (3.35)$$

This is the impulse response of the channel with low turbulence. The value of mean square amplitude fluctuation σ_X^2 is [15-16]:

$$\sigma_X^2 = 0.56\left(\frac{2\pi}{\lambda}\right)^{7/6}\int_0^L C_n^2(x)(L-x)^{5/6}\,dx \quad \text{for plane waves and} \qquad (3.36)$$

$$\sigma_X^2 = 0.56\left(\frac{2\pi}{\lambda}\right)^{7/6}\int_0^L C_n^2(x)\left(\frac{x}{L}\right)^{5/6}(L-x)^{5/6}\,dx \qquad (3.37)$$

for spherical waves where, L is the length of the channel. The average light intensity is:

$$\langle I\rangle = I_0\exp(2\sigma_X^2) \qquad (3.38)$$

The Negative Exponential fading model emerges from a scattering model that views the composite field as produced by a large number of non-dominating scatterers [16], each contributing to random optical phase at the detector. The central limit theorem then gives a complex Gaussian field, whose amplitude is Rayleigh. In this case, the random intensity is a one-sided exponential random variable, or chi-squared with two degrees of freedom. The complex field A, is *Rayleigh distributed* given as:

$$f_A(a) = 2a\exp(-a^2),\ a > 0 \qquad (3.39)$$

with normalization so that $\langle I\rangle = 1$. This implies negative exponential statistics for the intensity; or:

$$f(I) = \frac{1}{I_0}\exp\left(\frac{-I}{I_0}\right), I \geq 0 \qquad (3.40)$$

where, $I_0 = E[I]$ is the mean irradiance. The distribution of the log-normal and negative exponential are shown in Fig. 3.3.

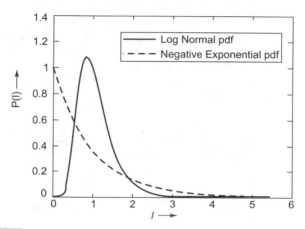

Figure 3.3 Log-Normal and Negative Exponential Distribution.

3.3.4.2 Gamma-Gamma PDF

In the case of Gamma-Gamma distribution, the irradiance of optical field is defined as the product of two random processes, i.e., $I = I_x I_y$, where I_x arises from larger turbulent eddies and I_y from smaller eddies. Specifically, gamma distribution is used to model both small-scale and large-scale fluctuations, leading to the gamma-gamma pdf in terms of irradiance I given by.

$$f(I) = \frac{2(\alpha\beta)^{(\alpha+\beta)/2}}{\Gamma(\alpha)\Gamma(\beta)} I^{(\alpha+\beta)/2} K_{\alpha-\beta}(2\sqrt{\alpha\beta I}) \tag{3.41}$$

where, Γ is the Gamma function and $K_{\alpha-\beta}$ is the modified Bessel function of the second kind of order $(\alpha-\beta)$. Here, α and β are the effective number of small-scale and large-scale eddies of the scattering environment [14]. These parameters can be directly related to atmospheric conditions for zero inner scale, which is relevant for upper bound calculations. For the plane wave case with aperture-averaged scintillation-index, the parameters are given by:

$$\alpha = \left[\exp\left\{ \frac{0.49\sigma_R^2}{(1+0.65d^2 + 1.11\sigma_R^{12/5})^{7/6}} \right\} - 1 \right]^{-1} \tag{3.42}$$

$$\beta = \left[\exp\left\{ \frac{0.51\sigma_R^2(1+0.69\sigma_R^{12/5})^{-5/6}}{(1+0.90d^2+0.62d^2\sigma_R^{12/5})} \right\} - 1 \right]^{-1} \tag{3.43}$$

For spherical wave propagation, these parameters are expressed as:

$$\alpha = \left[\exp\left\{ \frac{0.49\chi^2}{(1+0.18d^2+0.56\chi^{12/5})^{7/6}} \right\} - 1 \right]^{-1} \tag{3.44}$$

$$\beta = \left[\exp\left\{ \frac{0.51\chi^2(1+0.69\chi^{12/5})^{-5/6}}{(1+0.90d^2+0.62d^2\chi^{12/5})^{7/6}} \right\} - 1 \right]^{-1} \tag{3.45}$$

where, $\sigma_R^2 = 1.23C_n^2k^{7/6}L^{11/6}$ is the Rytov variance, $\chi^2 = 0.4\sigma_R^2$ is the Rytov's variance for spherical waves. The parameter $d = \sqrt{\frac{kD^2}{4L}}$, with D the diameter of the aperture. For $\sigma_R^2 \leq 1$ the gamma–gamma distribution resembles a log-normal distribution. As the turbulence strength increases, the distribution skews towards smaller values of irradiance and becomes more like negative exponential.

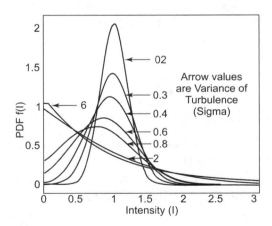

Figure 3.4   **Intensity Distribution at Different Variance Values of Turbulence Strengths for Gamma-Gamma Distribution.**

3.4 SYSTEM OPTICAL TRANSFER FUNCTION

The Optical wireless system OTF includes the transfer function of the optical components such as concentrators, lens system used in the transceivers and also that of the wireless channel to describe the channel model in terms of spatial frequencies as discussed in the earlier section.

3.4.1 The System's Optical Transfer Function

The system OTF is formulated as

$$\xi(f_x,f_y) = \xi_{(aer)}(f_x,f_y) \cdot \xi_{(tur)}(f_x,f_y) \cdot \xi_{(opt)}(f_x,f_y) \tag{3.46}$$

As indicated in (3.46) and Fig. 3.2, the system OTF is composed of the following components:

➢   optical hardware OTF; $(\xi_{(opt)}(f_x,f_y))$
➢   aerosol scattering OTF; $(\xi_{(aer)}(f_x,f_y))$
➢   turbulence OTF; $(\xi_{(tur)}(f_x,f_y))$

3.4.1.1 Diffraction-Limited Optics OTF

The optical lens system used in the transmitter and receiver to collimate and focus optical beams in wireless optical communication has significant effect on system performance. The diffraction-limited optics formulae refer to an equivalent lens of aperture diameter and focal length. As the diffraction-limited optics does not affect the position of the received signal, the OTF does not include the PTF component and consequently the MTF is given by [3-4],

$$\xi_{(opt)}(\omega,\lambda) = \frac{2}{\pi}\left[\cos^{-1}\left(\frac{\omega}{2\alpha_0}\right) - \frac{\omega}{2\alpha_0}\sqrt{\left(1-\frac{\omega}{2\alpha_0}\right)^2}\right] \cdots \omega \leq 2\alpha_0$$

$$0, \qquad\qquad \omega \geq 2\alpha_0 \tag{3.47}$$

where,

$$\alpha_0 = \frac{k_0 D}{2}, \omega = 2\pi f_r, f_r = \sqrt{(f_x^2 + f_y^2)}$$

where, D is the receiver-aperture diameter, f_r is the radial spatial frequency in cycles per meter.

3.4.1.2 Aerosol Scattering OTF

Aerosol in the atmosphere causes both absorption and scattering. The aerosol scattering OTF is, therefore, expressed in terms of the scattering and absorption coefficients Sa and Aa. The aerosol scattering OTF also does not affect the point position and thus contains only an MTF component and no PTF part. Hence, the MTF for aerosol scattering can be expressed as in [4]:

$$MTF_{(aero)}(f_a) = \left\{ \exp\left(-S_a L \left(\frac{f_a}{f_{ac}} \right)^2 \right) \right\}$$

$$. \exp\left\{ \left[\exp\left\{ -S_a L \left[1 - \left(\frac{f_a}{f_{ac}} \right)^2 \right] \right\} \right] \right\} (-A_a L) - \exp(-S_a L).$$

$$\text{for } f_a < f_{ac}$$

$$= \exp(-S_a L)\exp\{(1 - e^{S_a L})\} - A_a L. \quad \text{for } f_a \geq f_{ac} \qquad (3.48)$$

where, f_a is the radial spatial frequency expressed in cycles per radian and f_{ac} is the cut-off spatial frequency. The cut-off frequency f_{ac} is calculated according to the receiver's characteristics, such as its FOV, dynamic range and spatial bandwidth [17]. The aerosol scattering MTF for the short (10 msec) and long (1sec) exposures remains nearly equal. Here, we assume that equation (3.48) is applicable for the case of a very small (10 nsec) integration time, which also exists for optical-wireless communication.

3.4.1.3 Atmospheric Turbulence OTF

The OTF associated with atmospheric turbulence can be described by either a long or short exposure model. The detector integration time for optical-wireless communication are in the order of nanoseconds and less and hence is governed by short exposure model. As with the aerosol scattering OTF, the position of a point ultimately does not alter. Hence, the turbulence OTF does not include a phase component and only the MTF component exists. For short exposures of less than 10 ms, the transfer function [4] is:

$$MTF_{(tur,Sh)}(f_a) = MTF_{(tur,L)}(f_a)\exp\left[1 - \frac{1}{b}\left(\frac{f_a \lambda}{D} \right)^{1/3} \right] \qquad (3.49)$$

where, $b = 1$ for near field and 2 for far field.

The long exposure model integrates all the temporal variations in the vicinity of the blur in the focal plane. Thus, the turbulence OTF represents the spatial spread of the signal ray averaged over time, which determines power spread over the detector. The MTF for long exposure is expressed as:

$$MTF_{(tur,L)}(f_a) = \exp\left(-57.53\chi f_a^{5/3}\lambda^{-1/3}C_n^2 L\right) \tag{3.50}$$

where χ is wave-shape constant of 3/8 for spherical waves and 1 for planar waves.

3.5 INDOOR CHANNEL MODELING

The indoor wireless channel is usually modeled as a fixed channel, since it does not significantly change unless the transmitter, receiver, or objects in the room are moved by tens of centimeters. Noise in these systems is signal-independent Gaussian in nature. Similar to the outdoor channel, we can write the baseband indoor channel model as [20]:

$$y(t) = x(t)^*h(t) + w(t) \tag{3.51}$$

where, the "*" symbol is the convolution. The frequency response of channel impulse response $h(t)$ is given by $H(f) = \int\limits_{-\infty}^{\infty} h(t)\exp(-j2\pi ft)dt$, as given in the earlier section. Once again (3.51) is the base band linear model of the indoor channel which is equivalent to a linear filter. The average received optical power is given by:

$$P_r = H(0)P_t \tag{3.52}$$

where, $H(0)$ is the channel dc gain defined as: $H(0) = \int\limits_{-\infty}^{\infty} h(t)dt$.

As the frequency responses of the indoor infrared channels are relatively constant near the zero frequency, in most of the situations the most important quantity is the dc gain $H(0)$ to characterize the channel. In the following sections, therefore, the dc gain for the LOS, Non-LOS and the Diffused links are given.

3.5.1   DC Gain of the Indoor Channel

LOS link   The dc gain for the LOS links can be estimated reasonably accurately by considering only the direct propagation path for the different configuration. Consider the link geometry shown in Fig. 3.5(a). For the radiation pattern of the transmitter beam to be axially symmetrical, the radiant intensity is expressed as $P_t B(\phi)$ in (W/sr), where $B(\phi)$ is the beam pattern. The intensity of the light signal at the receiver, located at distance 'L' and at an angle ϕ with respect to the transmitter, is $P_t \dfrac{B(\phi)}{L^2}$ W/cm^2. The received power is then expressed as [20]:

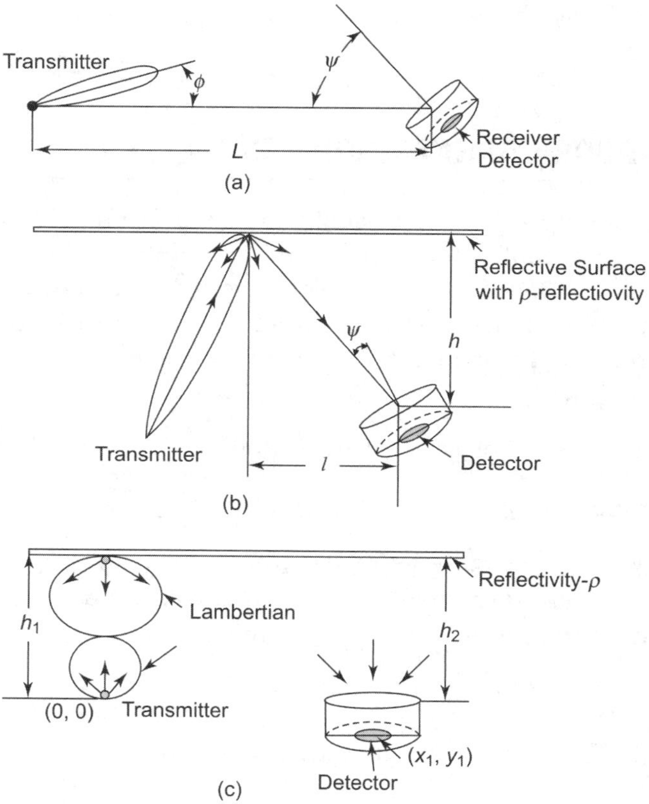

Figure 3.5 (a) Directed LOS (b) Hybrid Non-LOS (c) Diffused System

$$P_r = P_t \frac{B(\phi)}{L^2} A_{eff}(\psi) \tag{3.53}$$

where, the effective aperture of the receiver detector is,

$$A_{eff}(\psi) = \begin{cases} AT_s(\psi)g(\psi)\cos(\psi)..................0 \leq \psi \leq \psi_c \\ 0..\theta > \psi_c. \end{cases} \tag{3.54}$$

where, A is the physical aperture of the detector, ψ is the incidence angle with the receiver axis, $T_s(\psi)$ is the filter transmission at an angle ψ and $g(\psi)$ is the concentrator gain. The concentrators used in the indoor systems can be either imaging or non-imaging type. For an ideal nonimaging concentrator the gain can be expressed as [20]:

$$g(\psi) = \begin{cases} \dfrac{n^2}{\sin^2 \psi_c} \dotfill 0 \le \psi \le \psi_c \\ 0 \dotfill \psi > \psi_c \end{cases} \tag{3.55}$$

where, n is the refractive index and ψ_c is the FOV of the concentrator. A hemi-spherical concentrator with filter is shown in Fig. 3.6.

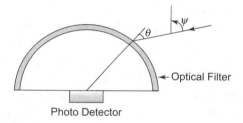

Photo Detector

Figure 3.6   Hemispherical Optical Concentrator with Filter.

From (3.53) and (3.54), the channel dc gain for the LOS link can then be expressed as:

$$H(0)_{LOS} = \begin{cases} \dfrac{A}{L^2} B(\phi) T_s(\psi) g(\psi) \cos(\psi) \dotfill 0 \le \psi \le \psi_c \\ 0 \dotfill \theta > \psi_c \end{cases} \tag{3.56}$$

From (3.56), besides the other factors, the dc channel gain can be increased by optimizing the transmitter radiant intensity $B(\phi)$. The transmitter beam pattern $B(\phi)$ for the practical LOS transmitters can be modeled well by using a generalized Lambertian radiation intensity, expressed as:

$$B(\phi) = \frac{m+1}{2\pi} \cos^m \phi \tag{3.57}$$

where, m is the order given by $m = \dfrac{-\ln 2}{\ln(\cos \phi_{1/2})}$ and $\phi_{1/2}$ is the transmitter semi-angle. Thus the beam or the angle $\phi_{1/2}$ gets narrower with increased value of the order m. The channel dc gain with a generalized Lambertian source is given by:

$$H(0)_{LOS,GEN} = \begin{cases} \dfrac{(m+1)A}{2\pi L^2}\cos^m(\phi)T_s(\psi)g(\psi)\cos\psi..............0 \leq \psi \leq \psi_c \\ 0...\theta > \psi_c \end{cases}$$ (3.58)

From the above expression, we observe that the dc gain for the generalized Lambertian source can be increased by narrowing the transmitter semi-angle $\phi_{1/2}$.

Non-LOS   In a Non-LOS configuration, also referred as hybrid non-LOS channels, the signals are received not only from the direct path, but also by the reflected light from walls and ceiling as shown in Fig. 3.5b. The diffuse reflectivity ρ, of the walls and ceiling lie in the range of 0.6–0.9 in the 800–900-nm range. Also, most building materials, except glass, are approximately Lambertian reflectors. To compute the gain of directed-non-LOS we assume that the transmitter illuminates a reasonably small spot at the ceiling. With this assumption the channel gain depends only on the horizontal separation between the illuminated spot and the receiver, and not on the separation between the transmitter and receiver. If the ceiling height is h above the receiver and l is the horizontal separation between the illuminated spot and the receiver, then the received signal irradiance is expressed as:

$$I_s(l,h) = \frac{\rho h P_t}{\pi(h^2 + l^2)^{3/2}}$$ (3.59)

For the received power of $P_r = I_s A_{eff}$, the channel dc gain of the hybrid-non-LOS configuration is given as [21]:

$$H(0)_{(D-H)non-LOS} = \begin{cases} \dfrac{\rho A h}{\pi(h^2 + l^2)^{3/2}}T_s(\psi)g(\psi)\cos(\psi)........0 \leq \psi \leq \psi_c \\ 0...\theta > \psi_c \end{cases}$$ (3.60)

The most effective means to increase $H(0)$ in this case are to increase the detector area and the concentrator gain, which can be increased by increasing the refractive index, and decreasing the FOV.

Non-directed-non-LOS   For the case of nondirected-non-LOS channels, also known as *diffused channels*, one must consider the effect of multiple reflections from different surfaces of the room, which are not specular but diffused in nature,. With reference to Fig. 3.5(c), as an approximation, we consider only the first reflection from the ceiling which is considered to be a large diffuser. In diffuse system the transmitter is directed straight upward and is assumed to emit a Lambertian pattern. The receiver also is pointing straight upward and is generally omni directional with a gain $g(\psi) \approx g = n^2$ and FOV of $\psi_c \approx \pi/2$. The filter is assumed to be having an omni directional pattern $T_s(\psi) = T_s$. Considering the transmitter and receiver to be located, respectively, at coordinates $(0, 0)$ and (x_1, y_1) in the horizontal (x, y) plane, we integrate the power reflected from each ceiling element to obtain the dc channel gain as [21]:

$$H(0)_{(ND)non-LOS} = \frac{\rho T_s g A h_1^2 h_2^2}{\pi^2} \iint \frac{dxdy}{(h_1^2 + x^2 + y^2)^2 \left[h_2^2 + (x - x_1)^2 + (y - y_1)^2 \right]^2}$$

(3.61)

For this configuration, the performance can be improved by increasing the detector area A, and the concentrator gain g, by increasing the refractive index n and not by reducing the FOV.

3.5.2 Modeling of Multi-path Diffuse Indoor Channel

In the above section, we have considered only single reflection in the non-directed systems. In practice, diffused infrared systems have multi-path propagation due to reflections from walls and ceiling. In this case, therefore, the channel characteristic will depend on the room size and shape. Multi-path propagation is usually dominated by diffuse reflections. To model the impulse response of the multi-path diffused channel different functional forms have been proposed [22], i.e., exponential decay model, ceiling-bounce model, etc. Certain statistical models, such as Raleigh, Gamma, etc., have also been used to model the impulse response much like in the case of outdoor links.

Considering the ceiling bounce model for the diffused multi-path channel response, the diffused reflection from a single infinite plane of the large ceiling is assumed to be a Lambertian. The functional model for the impulse response has one free parameter, which controls the channel delay spread or the temporal dispersion of the channel response due to multi-path. This functional model is suitable for the diffuse links consisting of a Lambertian transmitter that is directed towards a diffuse reflector of infinite extent, and with a receiver co-located with the transmitter. The response can then be given in closed form as [23]:

$$h(t) = \frac{H(0) \cdot 6\tau^6 \cdot u(t)}{(t+\tau)^7} \tag{3.62}$$

where, $u(t)$ is the unit step function and τ is the minimum time for the signal to travel from the transmitter and reach back the receiver after a reflection from the ceiling. This is related to the rms delay spread D_{rms} by:

$$\tau = \frac{2h}{c} = 12\sqrt{\frac{11}{13}} D_{rms} \tag{3.63}$$

Although this model is derived by considering an unshadowed diffuse link, it is found to be accurate for LOS and diffuse links with or without shadowing as well. The delay spread indicates the ISI induced by multi-path, This is computed from the impulse response using the following relation:

$$D_{rms} = sqrt \frac{\int\limits_{-\infty}^{\infty} (t-t_0)^2 h^2(t) dt}{\int\limits_{-\infty}^{\infty} h^2(t) dt} \tag{3.64}$$

where, the mean delay t_0 is given by:

$$t_0 = \frac{\int\limits_{-\infty}^{\infty} t h^2(t) dt}{\int\limits_{-\infty}^{\infty} h^2(t) dt} \tag{3.65}$$

As long as the positions of the transmitter, receiver and intervening reflectors are fixed, $h(t)$ and D_{rms} are fixed and hence are considered to be deterministic quantities.

The diffused link impulse response can also be simulated with good accuracy by simulating multi-path propagation of arbitrary number of diffuse reflections. The energy reflected from each surface element is assumed to follow a Lambertian distribution, independent of the angle of incidence. For the simulation of indoor multi-path infrared propagation, it is assumed that the light from the given source can reach the given receiver after any number of reflections from walls and ceiling with decreasing power after each reflection. Therefore, the impulse response can be written as an infinite sum [24]:

$$h(t, Tx, Rx) = \sum_{0}^{\infty} h^{(k)}(t, Tx, Rx) \qquad (3.66)$$

where, $h^{(k)}(t)$ is the response of the light undergoing k reflections between the transmitter Tx and receiver Rx. Starting from the direct response $h^{(0)}(t)$ between the transmitter and receiver each higher order term $h^{(k)}(t)$, $k>0$, is then calculated recursively from $h^{(k-1)}(t)$:

$$h^{(k)}(t, Tx, Rx) = \sum_{i=1}^{N} h^{(0)}(t, Tx, M_i) * h^{(k-1)}(t, M_i, Rx) \qquad (3.67)$$

where, M_i, $1 \geq i \geq N$ are the reflecting surfaces in the room. Equation (3.67) suggests that the k-bounce indoor channel models the impulse response by first finding the distribution of the reflectors M_i in the room due to the LOS response from the transmitter. Later using the M_i reflectors as the source, the $(k-1)$-bounce response of the receiver is obtained. The simulated k-bounce $h^{(k)}(t)$ has been shown in Fig. 3.7. for $k = 1 - 3$ values.

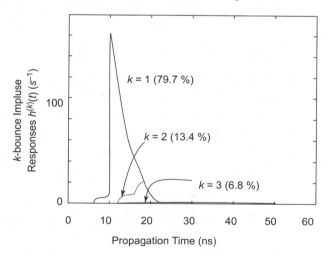

Figure 3.7 Simulated k-Bounce $h^{(k)}(t)$ for k = 1 to 3 Values [24].

3.6 MODELING NOISE SOURCES IN FSO

FSO links, both outdoor and indoor, are operated in the presence of intense infrared and visible background light. The ambient sunlight is the main source of the background radiation, which is detected as shot noise at the detector. Due to its high intensity, this shot noise is modeled as *white, Gaussian,* and *signal-independent.*

The indoor optical system, besides being affected by intense ambient infrared radiation, also has noise from the incandescent and fluorescent lamps, and other sources [25], [26]. Fluorescent lamps have spectral lines of mercury and argon that lie in the 780–950-nm band in the infrared systems. Besides these spectral lines, the fluorescent-lamp electrical power spectrum contains discrete components at harmonics of the drive frequency of 50 or 60 Hz. These harmonics are present with significant power up to tens of kilohertz. On the other hand, the electronic ballasts drive the lamps at frequencies of tens to hundreds of kilohertz. Their detected electrical spectrum contains energy up to hundreds of kilohertz [27-28].

All these noise sources produce signal-independent shot noise in the detected DC photocurrent. If the ambient light originates from a source at angle ψ with respect to the receiver normal, then the received ambient optical average power is [21] given as:

$$P_n = p_n \Delta\lambda_n T_0 Ag(\Psi)\cos\Psi \tag{3.68}$$

where, $\Delta\lambda_n$ is the noise bandwidth of the bandpass optical filter with peak transmission coefficient T_0, p_n is the constant spectral irradiance of the ambient light noise and A is the detector area. This in turn produces a noise current in the receiver as:

$$w(t) = \Re P_n \tag{3.69}$$

Summary

The signal is affected by the atmospheric channel by the large-scale variation in the signal strength due to atmospheric effects, such as, fog, rain, etc. The small-scale fading is the variation of signal strength due to atmospheric turbulence and interference due to multi-paths. From system theory perspective, a channel is modeled as a linear system, represented as a input-output baseband system with an impulse response $h(t)$. The atmosphere is characterized by key parameters: spectral spread and Delay spread. The spectral spread, Ds, is related to the coherence time and is proportional to the angular spread of the arriving paths. Delay spread, Td, which is related to the coherence bandwidth, is proportional to the difference between the lengths of the shortest and the longest paths. The impulse response for the random turbulent atmospheric channel has different statistical channel models: Log-Normal, Exponential, Gamma-Gamma, etc. The indoor channel impulse response for the case of LOS-directed and the diffused system are considered to be deterministic. The noise in the atmosphere is considered to be additive. The ambient noise due to solar irradiance in the atmospheric channel is modeled as white Gaussian and signal-independent. In the case of indoor systems, besides the noise due to ambient light, which is white Gaussian signal independent, there are also discrete noise components at harmonics of 50 or 60 Hz, upto tens of kilohertz and upto hundreds of kilohertz due to fluorescent lamps and electronic ballasts drive current, respectively.

References

1. Pahlavan K., 'Wireless communication for office Information Network', *IEEE Communication Magazine*, 23(6), 19–27, 1985.
2. Proakis John G., Salehi M., *Fundamentals of Communication Systems*, Pearson Education, 2nd ed., 2008.
3. Williams C.S. and Becklund O.A., *Introduction to the Optical Transfer Function*. New York, Wiley,1989.
4. Kopeka N.S., *A System Engineering Approach to Imaging*, Bellingham, WA, SPIE, 1998.
5. Sasiela, R.J., *Electromagnetic Wave Propagation in Turbulence*, New York, Springer Verlag, 1994.
6. Ishimaru A., *Wave Propagation and Scattering in Random Media*, New York, Academic, 1978, Vol 1–2.
7. Andrews L.C. and Phillips R.L., *Laser Beam Propagation through Random Media*, 2nd ed., SPIE Press 2005.
8. Zhu X., Kahn J. M., 'Communication techniques and coding for atmospheric turbulence channels', *J. Opt., Fiber Communications*, 363–405 (2007), 2007 Springer Science.
9. Hufnagel R.E. and Stanley N.R., 'Modulation transfer function associated with image transmission through turbulent media', *Journal of Optical Society. America*, Vol. 54, pp. 52–61, Jan 1964.
10. Valley G.C., "Isoplanatic degradation of tilt correction and short-term imaging system," *Appl. Opt.* **19**, 574–577, (1980).
11. Goodman J.W., *Statistical Optics*, New York: Wiley 1985.
12. Andrews L.C., Phillips R.L., 'Mathematical genesis of the I-K distribution for random optical fields', *Journal of Optical Society America A*, Vol. 3, no.11, pp. 1912–1919. Nov.–Oct. 1986.
13. Churnside T.H. and Clifford S.F., 'Log-normal Rician probability density function of optical scintillations in the turbulent atmosphere', *Journal of Optical Society America A*, Vol 4, no. 10 pp. 1923–1930. Oct. 1987.

14. AL-Habash M.A., Andrews L.C., Phillips R.L., 'Mathematical model for irradiance PDF of a laser propagating through turbulent media', *Opt. Eng.* Vol. 40, No. 8, 2001, pp. 1554–1562.
15. Andrews L.C., Phillips R.L. and Hopen C.Y., *Laser Beam Scintillation with Applications*, SPIE Press, 2001.
16. Karp S., Gagliardi R., Moran S.E. and Stotts L.B., *Optical Channels*, New York, Plenum, 1988.
17. Bushuev D., Kedar D. and Arnon S., 'Analysing the Performance of a Nanosatellite Cluster-Detector Array Receiver for Laser Communication', *Journal of Lightwave Technology*, Vol. 21, No.2, Feb. 2003, pp 447-455.
18. Amon S., 'Optical Wireless Communication', in *The Encyclopedia of Optical Engineering*, New York: Marcel-Dekker
19. Bushnev D., and Arnon S., 'Analysis of the performance of a wireless optical multi-input to multi-output communication system', *Journal of Optical Society America. A*, Vol 23, No 7, July 2006, pp. 1722–1730.
20. Savicki J.P. and Morgan S.P., 'Hemispherical concentrators and special filters for planar sensors in diffuse radiation fields,' *Appl. Optics*, Vol. 33, no. 34, pp 8057–8061.
21. Kahn J.M. and Barry J.R., 'Wireless Infrared Communication', Proceedings of the IEEE, 85(2), 265-298, 1997.
22. Perez-J R., Berges J., Betancor M.J., 'Statistical model for the impulse response on infrared indoor channels, *Electronics Letters*, Vol. 33, no 15, 1997.
23. Carruthers J.B., Kahn J.M.,' Modeling of nondirected wireless infrared channels', *IEEE Transactions on Communications*, Vol. 45, no.10, Oct. 1997.
24. Barry J.R. and Kahn J.M., Krause W.J., Lee E.A. and Messerschmitt D.G. 'Simulation of multi-path impulse response for indoor wireless optical channels,' *IEEE Journal on Selected Area of Communications*, Vol. 11, 1993, pp 367–379.
25. Boucouvalas A.C., "Indoor ambient light noise and its effect on wireless optical links", *IEE Proceedings-Optoelectronics*, Vol. 143, 1996, pp 334–338.
26. Barry J.R., *Wireless Infrared Communication*, Kluwer Academic Publishers, 1994.
27. Narasimhan R., M.D. Audeh, and J.M. Kahn, 'Effect of electronic-ballast fluorescent lighting on wireless infrared links', *IEE Proceedings Optoelectronics*, Vol. 143, pp 347–354, 1996.
28. Moreira A., Valadas R., and Duarte A., 'Performance of Infrared transmission systems under ambient light interference', *IEE Proceedings Optoelecronics*, Vol. 143, no. 6, Dec. 1996, pp. 339–346.

Chapter

Modulation Techniques

The information signal can be transmitted through the optical wireless channel using one of the several modulation techniques, which are normally used in the optical fiber communication. The simplest method to modulate the light is the direct intensity modulation of the laser beam by base-band information signal. Alternatively, the base-band signal can be translated onto an electrical RF sub-carrier by means of one of the analog or digital modulation scheme prior to intensity modulation of the optical source.

In the present chapter, we study the various modulation schemes that are commonly used in FSO Systems. Starting with constraints of the FSO channel in Section 4.1, in Section 4.2, we classify the modulation schemes used in FSO Systems. In Section 4.3, the criteria applied for the selection of one scheme over the other is mentioned. Then in Section 4.4, we go on to give the vector channel modeling and optimum detection of the FSO in order to know the performance of these modulation schemes. In Section 4.5, expressions for minimum power requirement and spectral bandwidth are derived. In the final Section 4.6, we discuss in detail the performance of the different modulation schemes used in FSO Systems.

4.1 CONSTRAINTS OF THE CHANNEL

It is not always possible to apply all modulation schemes used in the RF wireless or the optical fiber system directly to wireless optical intensity channels. There are certain constraints of the FSO channel mentioned in Chapter 3, which have to be taken into account when selecting the modulation schemes. These are:

(i)   The transmitted optical intensity signal, which is modulated by the information signal must remain non-negative for all times since the transmitted power can physically never be negative, i.e., $P_t(t) \geq 0$. This constraints the electrical information signal to be non-negative as well, i.e., $x(t) \geq 0$.

(ii)   For the non-coherent IM-DD systems, which are normally used in the terrestrial and the indoor systems, the information is only in the intensity of the transmitted signal.

(iii) The average optical power is limited to some fixed value P, which satisfies eye/skin safety regulations, $P_t(t) = \lim_{T \to \infty} \dfrac{1}{2T} \int\limits_{-T}^{T} P_t(t)dt \leq P$ or in other words, $\lim_{T \to \infty} \dfrac{1}{2T} \int\limits_{-T}^{T} x(t)dt \leq P$. This is in contrast to conventional RF channels, in which the average square amplitude of the transmitted signal is constrained.

4.2 TYPES OF MODULATION SCHEMES

As mentioned above, in IM-DD optical wireless systems, the modulation techniques can be grouped into two general categories: base-band intensity modulation and sub-carrier IM modulation. In base-band modulation, at the optical transmitter end, the information signal directly modulates the LED/LD drive current of the optical carrier. At the receiver end, the information is recovered using the technique of direct detection, in which the photo detector generates an electrical signal according to the instantaneous power of the received optical signal. In the case of sub-carrier modulation, the information signal first modulates the RF electrical sub-carrier. The modulation scheme can be BPSK, QAM, AM, FM etc. This modulated electrical sub-carrier signal in turn intensity-modulates the optical carrier. At the receiver end the signal is recovered once again by direct detection. All these techniques are non-coherent which are adopted due to low cost and lower complexity of the receiver structure used in the terrestrial and indoor systems. However, some of the space laser communication links do use coherent schemes as well with better performance, but with added complexity in the transreceiver structures.

A number of digital, analog and pulse modulation techniques can be used in the optical wireless schemes. Among the *analog schemes*, AM, FM, and PM can be used [1], more so in indoor application and for video signals. In *pulse modulation*, sequence of carrier pulses with a suitable parameter, such as the pulse amplitude, width or position is electrically modulated by the base-band as PAM, PWM or PPM modulation, respectively. Again, the modulated carrier is transmitted optically by intensity modulation of the optical signal. In *digital schemes*, which are most commonly used because of their inherent advantage, prior to transmission the information is digitized and translated to a specific code, such as RZ, or NRZ codes to get a stream of pulses which are then modulated with one of the digital schemes, i.e., OOK, PPM, DPSK, etc.

4.3 SELECTION CRITERIA

Selecting the more appropriate modulation scheme will depend on certain system requirement criteria. For optical wireless systems, the two main criteria are:

➤ *Power efficiency*

➤ *Bandwidth efficiency*

To study the performance of any modulation format, the important parameters are its bandwidth and the minimum power requirement at the receiver detector to correctly detect the signal in the presence of noise. As the average transmitted optical power governs the eye safety and electrical consumption of the transmitter, it is essential to calculate the power requirement for each modulation scheme. The bandwidth requirement on the other hand is important as it determines the maximum data rate achievable by the link with a particular modulation format.

Power-efficient modulation schemes are simple to implement, effective in compensating for the channel path loss for low data rate as in the case of LOS indoor link. The modulation scheme with short pulses, such as, PPM meets the low average power requirement essential for the eye safety consideration, but they require a larger bandwidth. On the other hand, at high data rate terrestrial links and for diffused indoor links, bandwidth efficient schemes, such as OOK, etc., are more effective. The diffused links have a bandwidth limited channel characteristic due to multi-path dispersion caused by reflections of light from walls and other objects in the room. Also, at high data rate in terrestrial links ISI is particularly pronounced because of atmospheric turbulence. Therefore, power efficient modulation schemes which may not be bandwidth efficient, such as PPM can not give good performance in these systems. On the other hand, in bandwidth efficient schemes such as OOK, the restrictions on power levels due to laser eye safety regulations significantly reduce the link margin, thereby restricting the operational range. In this regard, we need to select power efficient modulation schemes, like PPM and its variants at some cost of bandwidth in order to enhance link performance at same SNR values. Therefore, a compromise has to be made between the power and bandwidth when selecting a modulation scheme.

4.4 VECTOR CHANNEL MODELING AND OPTIMUM DETECTION

4.4.1 Vector Channel Model

The detection of the received optical signal is done in the electrical domain, therefore, in order to analyze the different modulation schemes and obtain their performance characteristics, we can use geometrical representation of signal waveforms as vectors in electrical signal space with the constraints as discussed above. The channel is also represented geometrically in this linear space as a vector channel. This representation provides a compact characterization of the communication link for the transmitting digital information.

Figure 4.1 gives the system model of FSO for digital signaling and with additive white Gaussian noise (AWGN). The information source sends one of the M information signal waveforms $s_m(t)$ from a set {S}; $1 \leq m \leq M$. Each of this electrical waveform is uniquely mapped to an optical intensity signal $x_m(t)$ and transmitted on the optical channel by the transmitter. Let all these optical intensity signals belong to a set {X}; $\{x_m(t):m \in M\}$. As here we are interested to determine the performance of the modulation schemes we assume an *ideal flat channel* with infinite bandwidth and no atmospheric turbulence. For signaling we assume an

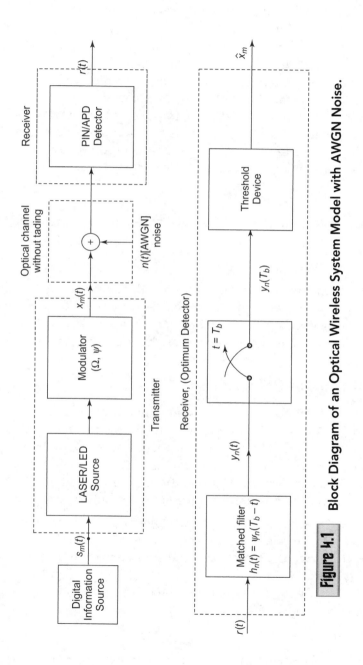

Figure 4.1 Block Diagram of an Optical Wireless System Model with AWGN Noise.

ideal pulse train with no overlap and a bit period T_b. These independently and identically distributed (i.i.d) signals can be written as:

$$x(t) = \sum_{k=-\infty}^{\infty} x_k(t - kT_b) \tag{4.1}$$

The signal space for the transmitted signal is obtained by specifying a set of $N \leq M$ orthonormal functions $\psi = \{\psi_n(t) : n \in N\}$ in a vector space V [2]. The orthonormal waveforms $\{\psi_n(t)\}$ form an orthonormal basis set in the N-dimensional signal space. The basis functions are energy functions normalized to unit energy. It is then possible to represent each $x_m(t)$ in the set $\{X\}$ as a linear combination of the elements of ψ as:

$$x_m(t) = \sum_{n \in N} x_{m,n} \psi_n(t) \tag{4.2}$$

In (4.2) the transmitted signal vector is represented in terms of the real coordinates as:

$$x_m = (x_{m,1}, x_{m,2}, \ldots x_{m,N}) \tag{4.3}$$

The dimensionality N of the signal space will be equal to M if all the M signal waveforms are linearly independent. Any modulation scheme can, therefore, be described by the pair (Ω, ψ), where Ω is the *signal constellation* of the modulated signal or collection of all the possible signal vectors.

4.4.2   Optimum Detection

In order to determine the power and bandwidth efficiency of the transmitted modulated signal we need to detect the received signal optimally. By optimum detection, we mean maximization of the sampled signal-to-noise ratio as well as providing sufficient statistics for the transmitted signal. Therefore, next we describe the signal detection at the receiver to recover the transmitted information optimally.

As discussed in Chapter 3, the optical channel can be represented by a baseband electrical model with the amplitude and power constraints on the transmitting signal. Hence, after transmission through the channel with a unity impulse function (i.e., assuming infinite bandwidth and flat response) and with AWG noise, the received baseband signal after detection can be expressed as:

$$r(t) = x(t) + w(t) \tag{4.4}$$

where, $w(t)$ is a Gaussian random process with mean zero and variance of $N_0/2$. N_0 is the double-sided noise power spectral density of the zero mean white Gaussian noise. The received signal is first detected and then it is demodulated. Based on the observation of received signal $r(t)$, over the symbol interval we process the detected signal that is optimum, in the sense that it minimizes the probability of making an error. The demodulator converts the received signal to a vector with the same dimension as that of the transmitted signal. The optimum detector is a bank of N matched filters with impulse response $h_n(t)$ *matched* to the basis function:

$$h_n(t) = \psi_n(-t) \tag{4.5}$$

The detected signal output of the matched filter detector at $t = T_b$ is:

$$y_n = y_n(T_b) = \int_0^{T_b} r(\tau) h_n(T_b - \tau) d\tau \tag{4.6}$$

or,

$$y_n = y_n(T_b) = \int_0^{T_b} [x_m(t) + w(t)] \psi_n(\tau) d\tau \tag{4.7}$$

As we can see from equation (4.7), output of the matched filter is basically the time-autocorrelation function of the transmitted signal and the noise term. Equation (4.7) hence reduces to:

$$y_n(T_b) = x_{mn} + w_n(T_b) \tag{4.8}$$

where, $x_{mn} = \int_0^{T_b} x_m(t) \psi_n(t) dt$; $n = 1,2,3,...,N$, $w_n(T_b)$ represents the AWGN noise component of the output signal at the sampling interval of bit time. We conclude, therefore from equation (4.8) that for a given vector signal transmission x_m, the output of the optimum detector matched filter will also be Gaussian random variable with mean x_m and variance of Gaussian random noise N_0.

4.5 MINIMUM POWER AND SPECTRAL BANDWIDTH

4.5.1 Minimum Power Requirement

The minimum power of the received detected signal is directly related to the expected error in the estimated signal at the receiver, or in other words, for a given Bit Error Rate (BER) the required minimum power of the received signal gets decided. Hence, we need to determine the relationship between received minimum power and BER to obtain the power efficiency of the modulated scheme.

As the detected signal $\bar{y}$ is Gaussian random signal, an estimate of the sent symbol $\hat{x}_m$ is obtained on the basis of the conditional probability density function of receiving signal $\bar{y}$ when signal x_m is transmitted [2]:

$$f(\bar{y}|x_m) = \frac{1}{(2\pi N_0/2)^{1/2}} \exp\left(-\frac{\|\bar{y} - x_m\|^2}{2N_0/2}\right) \tag{4.9}$$

Since the components of the noise are independent and all have the same mean and variance, the distribution of the noise vector $w_n(T_b)$ in the N-dimensional signal space has spherical symmetry. The received vector $\bar{y}_n$ in the signal space can then be represented by a spherical cloud centered at the transmitted signal x_m or s_m. The density of the noise cloud around the deterministic signal x_m is dependent on the variance of the noise. The larger spread of the cloud indicates higher probability of error and vice-versa. The detector, which makes the decision for least probability of error or probability of making maximum correct decision, is based on the computation of $P(x_m|\bar{y})$; the *maximum â-posterior probability* (MAP), defined as the probability of transmitting x_m when $\bar{y}$ is received. Using Bayes's rule, we may express the posterior probability as:

$$P(x_m|\bar{y}) = \frac{f(\bar{y}|x_m)P(x_m)}{f(\bar{y})} \tag{4.10}$$

where, $P(x_m)$ is the *â*-priori probability of the mth signal before being transmitted and $f(\bar{y})$ is expressed as:

$$f(\bar{y}) = \sum_{m=1}^{M} f(\bar{y}|x_m)P(x_m) \tag{4.11}$$

In the case, when the source selects the symbols with equal probability, or when $P(x_m) = 1/M$ for all M, which makes $f(\bar{y})$ independent of the signal being transmitted, the MAP condition given by equation (4.10) reduces to finding the signal which maximizes $f(\bar{y}|x_m)$. This is called the *maximum likelihood* function and the detector based on the maximum of $f(\bar{y}|x_m)$ over M signals is called the *maximum likelihood (ML) detector*. This chooses $\hat{x}_m$ to maximize the posterior probability in equation (4.10). In other words, the ML detector selects $\hat{x}_m$, such that it minimizes the Euclidean distance in $\|\bar{y} - x_m\|^2$ in (4.9) [3].

Probability of symbol error is an important parameter to determine the performance of any modulating schemes in the AWGN channel. This can be defined as; $P_{sym} = \Pr(\hat{x}_m \neq x_m | x_m sent)$ or from equation (4.9) as:

$$P(\bar{y}|x_m) = \int_y f(\bar{y}|x_m)dy = \frac{1}{\sqrt{\pi N_0}} \int_y \exp\left(-\frac{|\bar{y} - x_m|^2}{N_0}\right)dy \quad (4.12)$$

It is difficult to calculate the symbol error probability exactly for different modulation schemes; therefore, bounds are used to approximate its value. There is a simple upper bound to the average value of symbol error probability of general equiprobable M-ary signaling schemes. This upper bound is known as the *union bound* [3, 4], given by:

$$P_e(sym) \approx N_{av}Q\left(\frac{d_{min}}{2\sqrt{N_0/2}}\right) \quad (4.13)$$

where, d_{min} is the minimum Euclidean distance between any pair of valid modulation signals or the square root of the minimum received signal, N_{av} are the average number of symbol points, which are d_{min} away from any other point in the constellation and $Q(.)$ is the area under the tail of the Gaussian pdf, defined as:

$$Q(x) = \frac{1}{\sqrt{2\pi}} \int_x^\infty e^{-u^2/2}du \quad (4.14)$$

We can, therefore, determine the minimum received power for the given symbol error from (4.13). This enables to determine the minimum power requirement at the receiver for a particular link length and to achieve the desired SNR. It is also desirable to convert the probability of a symbol error to an equivalent probability of bit error in order to compare the various modulation schemes. It is not possible to determine this exactly in all cases. For the case of equiprobable orthogonal signals the average bit-error probability for a k-bit symbol is:

$$P_e(b) = \frac{2^{k-1}}{2^k - 1} P_e(sym) \quad (4.15)$$

4.5.2 Spectral Characteristics

The spectral characteristic of a modulated signal helps to determine the bandwidth requirement of the system for the particular modulation scheme. The bandwidth occupied by a modulation scheme indicates the spectral support required for the undistorted transmission of the modulated signal and hence the maximum data rate. The *bandwidth efficiency* of a modulation scheme, used as a figure of merit for the comparison among the schemes, is defined as the ratio of the bit rate R_b and the required bandwidth. Bandwidth of the modulation signal can be defined as 3 db or also as the first null bandwidth.

Assuming the transmitted signal to be wide-sense cyclostationary random process we determine the distribution of power over the different frequency components by the *power spectral density (PSD)* $S_X(\omega)$, of the modulated signal. This has two parts; discrete and continuous spectrum. The discrete part is the periodic component of clock rate which consists of discrete components of frequency at the bit rate and its harmonics. This is required for the clock recovery at the receiver. It does not carry any information but does require energy. The DC spectral component gives the average power of the optical signal in the modulation scheme. The continuous spectrum on the other hand has the information of data and the pulse shape. The PSD of the digitally modulated optical time disjoint pulse train described by (4.1) for independent and equally likely signaling is expressed as [3, 5]:

$$S_X(\omega) = \frac{1}{T}|P(\omega)|^2 S_{Xd}(\omega)$$

where, $P(\omega)$ is the Fourier transform of the rectangular pulse $x(t)$ with $|P(\omega)|^2 = \operatorname{sin}c^2\left(\dfrac{\omega T}{2\pi}\right)$ is the continuous spectra with $\operatorname{sin}c(x) \equiv \dfrac{\sin(\pi x)}{\pi x}$.

$S_{Xd}(\omega)$ is the PSD of the pulse sequence. It is the series of Dirac delta functions in the spectrum. $S_{Xd}(\omega)$ can be obtained according to the *Wiener-Kinchine theorem* by the Fourier transform of the autocorrelation function of the pulse sequence of the particular digital modulation scheme. For the random pulse sequence, $S_{Xd}(\omega)$ is expressed as:

$$S_{Xd}(\omega) = 1 + \frac{1}{T}\sum_{m=-\infty}^{\infty}\delta\left(\omega - \frac{m}{T}\right)$$

Hence, the continuous and discrete components of $S_X(\omega)$ are now explicitly expressed as:

$$S_X(\omega) = \frac{1}{T}|P(\omega)|^2 + \frac{1}{T^2}\sum_{m=-\infty}^{\infty}\left|P\left(\frac{m}{T}\right)\right|^2 \delta\left(\omega - \frac{m}{T}\right) \tag{4.16}$$

As we have modeled the optical channel by its electrical base-band equivalent with constraint on the signal level to be non-negative and of limited power; therefore, the bandwidth of the electrical PSD will be appropriate to obtain correctly the bandwidth of the optical spectrum. Also, the spectral response can be tailored accordingly for the optical channel by appropriately designing the modulation scheme and the pulse shape.

4.6   MODULATION SCHEMES USED IN FSO SYSTEMS

4.6.1   On-Off Keying

On-Off Keying (OOK) is one of the most common modulation schemes used in the optical wireless systems. The transceiver structures used are simple and the modulation scheme has high bandwidth efficiency. It is a binary level scheme with two symbols. The presence of an optical carrier wave indicates a binary 'one' symbol and its absence indicate a 'zero' symbol. Most of the other more advanced but efficient modulations can be expressed and studied in terms of OOK. The OOK optical signal in terms of its basis function can be expressed as:

$$x(t) = \sum_{k=-\infty}^{\infty}\sqrt{T_b}a[k]\psi_{OOK}(t-kT_b) \tag{4.17}$$

where, $a[k] \in \{2P,0\}$ and are chosen uniformly with a constant value of optical peak power of 2P and 0 for 'one' and 'zero', respectively for the two symbols. The basis function is defined as:

$$\psi_{OOK}(t) = \frac{1}{\sqrt{T_b}}rect\left(\frac{t}{T_b}\right) \tag{4.18}$$

where,

$$rect(t) = \begin{cases} 1 & : \quad 0 \leq t \leq 1 \\ 0 & : \quad \text{otherwise} \end{cases}$$

$x(t)$ satisfies the non-negative criterion, as the basis functions are non-negative in the symbol period. The OOK basis functions and the constellation of the two OOK signals are shown in Fig. 4.2. These are two points in the single dimension signal space.

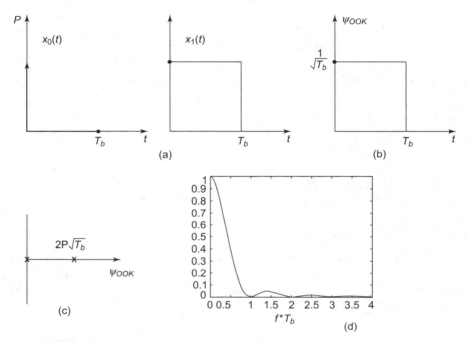

(a)   (b)

(c)   (d)

Figure 4.2   OOK Modulation (a) NRZ Symbol Set; $x_0(t)$ and $x_1(t)$: '0' and '1' Symbol (b) Basis Function for OOK, (c) Constellation Diagram (d) Power Spectral Density of the Continuous Part of OOK Spectrum for $P=1$ and $T_b=1$.

The bandwidth required by OOK can be determined from the continuous part of power spectral density (4.16) of the OOK signal as:

$$S_{OOK}(f) = P^2 T_b \operatorname{sin} c^2(\pi f T_b) \qquad (4.19)$$

The continuous power spectrum is shown in Fig. 4.2(d). With respect to first null, the bandwidth for OOK is $BW_{OOK} = R_b = 1/T_b$, the inverse of the pulse width. The bandwidth efficiency, therefore, is unity $\left(= \dfrac{R_b}{BW} \right)$ for NRZ OOK modulation.

Next, assuming matched filtering with maximum likelihood (ML) receiver the average probability of symbol error of an OOK system from (4.13) is:

$$P_e(sym) = Q\left(\frac{d_{min}}{2\sqrt{N_0/2}}\right) = Q\left\{\frac{P_{min}}{\sqrt{R_b N_0/2}}\right\} \tag{4.20}$$

where, N_0 is the double-sided noise power spectral density of the zero mean white Gaussian noise, P_{min} is the minimum average received optical power in (watts) at the detector. The minimum Euclidean distance between the two signals in the OOK signal set is:

$$d_{OOK} = \frac{2P_{min}}{\sqrt{R_b}} \tag{4.21}$$

Since the symbol rate is same as the bit rate, the symbol error is also equal to the bit error. By inverting equation (4.20), the minimum power requirement for OOK for given bit error is obtained as:

$$P_{min} = \sqrt{R_b N_0/2} . Q^{-1}[P_e(b)] \tag{4.22}$$

OOK is a bandwidth efficient modulation scheme and is, therefore, useful at high data rates, say 100 Mbps or greater. But it has reduced power efficiency. To reduce the average power requirement the pulse can have different duty cycles. By using a duty cycle less than 1, the required bandwidth though is increased by a factor of $1/d$ (d is the duty cycle) but the average power requirement can be reduced.

The OOK modulator transmitter emits rectangular pulses of duration $1/R_b$ and peak intensity of $2P$ to generate bit 'one', and no pulse to generate bit 'zero'. For the detection of OOK signals in the absence of distortion, but in the presence of AWGN, the ideal maximum-likelihood demodulator is a continuous-time matched filter, which is matched to the pulse shape. The output of the demodulator is sampled and estimated by threshold detection [2]. In the presence of multi-path phenomenon other types of filters are employed, such as a Whitened Matched Filter (WMF) performing maximum-likelihood sequence detection (MLSD) [6].

Restriction on laser power output, due to safety concerns to human eye and skin, requires power efficient modulation schemes for better range. In FSO communications fog, scintillation and scattering can lead to a decrease in the power density of the transmitted beam, limiting the range of the link. Therefore, though OOK is a popular choice in indoor systems, for outdoor systems higher average power efficient modulation schemes, i.e., PPM, PAM, DPPM have become important to convey information. Proper choice of these modulation techniques is very important with respect to the link design keeping in view the various parameters of interest that govern its performance. With the OOK modulation scheme as a benchmark, comparison among different modulation schemes is studied further.

4.6.2   M-Pulse Position Modulation (M-PPM)

Digital Pulse Position Modulation PPM [7] is considered to be one of the best modulation techniques for power-limited intensity modulation with direct detection communication systems. PPM has been widely used in terrestrial optical communication systems, and has also been adopted by the IEEE 802.11 for the infrared physical layer standard [8-9].

M-PPM is a form of orthogonal signaling defined to have M slots in a single symbol time using two intensity levels. These M-slots are called the 'chips'. A block of input bits is mapped to one of M distinct waveforms, each including one 'ON' chip and (M-1) 'OFF" chips. The symbol intervals have a duration given by:

$$T_b = \frac{\log_2 M}{R_c} \tag{4.23}$$

where, R_c is the chip rate.

As the transmitter emits an optical pulse during only one of the chips, which has duration T_b/M, in an M-PPM a constant peak optical power of MP watts is transmitted within only one of chips while the remaining (M-1) chips will have zero power. Here P is average transmitted power. As the chips are non-overlapping, therefore, each symbol is orthogonal. The signal space of M-PPM is an M dimensional Euclidean space with single point on each of the M-axis. The time domain representation of the intensity waveform in term of the basis functions is:

$$x(t) = \sum_{k=-\infty}^{\infty} \sqrt{\frac{T_b}{M}} MP\psi_{[k]M-PPM}(t - kT_b) \tag{4.24}$$

where, k will select a symbol equiprobably and uniformly from M, and the basis function ψ_{M-PPM}, for $m \in M$ will take the form as:

$$\psi_{(m)M-PPM}(t) = \sqrt{\frac{M}{T_b}} rect\left(\frac{1}{T_b/M}\left[t - (m-1)\frac{T_b}{M} \right] \right) = 1 \text{ for } (m-1)\frac{T_b}{M} \leq t \leq m\frac{T_b}{M} \tag{4.25}$$

$$= 0 \text{ otherwise}$$

We can see from (4.24) and (4.25) that for $M=1$, PPM reduces to OOK waveforms. The pulses are always non-negative due to the basis function selection and equally probable in different symbols. Peak power in each symbol is MP to restrict the average power to P. Figure 4.3 shows the waveforms for binary 4-PPM signals.

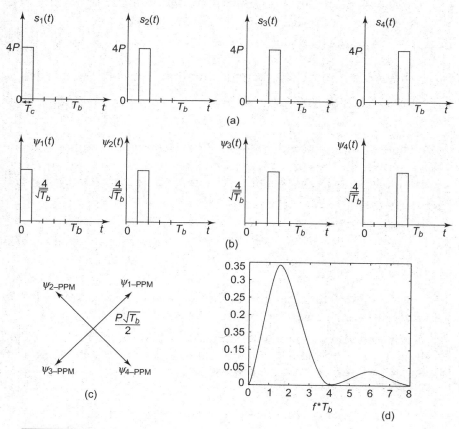

Figure 4.3 M-PPM Modulation (a) 4-PPM Symbols (b) Basis Functions for 4-PPM (c) Constellation Diagram (d) PSD of 4-PPM for $P=1$ and $T_b=1$.

The bandwidth for PPM is once again calculated from the continuous part of the *PSD*, given by [10]:

$$S_{M-PPM}(f) = P^2 T_b \sin c^2\left(\frac{\pi f T_b}{M}\right)\left[1 - \frac{1}{M^2}\left\{M + 2\sum_{i=1}^{M-1}(M-i)\cos\left(\frac{2\pi f T_b}{M}i\right)\right\}\right]$$

(4.26)

From above, the first null bandwidth for M-PPM is:

$$BW_{M-PPM} = \frac{M}{T_b}$$

(4.27)

The bandwidth efficiency, therefore, for first null can be written as:

$$\frac{R_c}{BW_{M-PPM}} = \frac{1}{M}\log_2 M \; \frac{\text{bits/sec}}{\text{Hz}}$$

(4.28)

The bandwidth requirement for M-PPM increases as compared to the OOK bandwidth. The chip rate is M times symbol rate R_b, and $\log_2 M$ bits are mapped in the block. Therefore, by simple calculation the bandwidth of M-PPM can be expressed as:

$$BW_{M-PPM} = \frac{M}{\log_2 M}.BW_{OOK}$$

(4.29)

Bandwidth for M-PPM, therefore, increases with M; for 4-PPM twice and for 16-PPM 4 times more bandwidth is required as compared to OOK. Figure 4.4 (d) gives the spectrum for 4-PPM.

As mentioned above, in the signal space, each PPM signal is orthogonal to all others signals and also equidistant to all the others. Based on this, the symbol error with AWGN is given as [2]:

$$P_e(sym) \approx (M-1)\cdot Q\left(P_{M-PPM}\sqrt{\frac{M}{2R_b N_0/2}}\right)$$

(4.30)

The bit error probability, due to equi-probability of symbols and their orthogonality in the signal space, can be expressed as [4]:

$$P_e(b) \approx \frac{M}{2}Q\left[P_{M-PPM}sqrt\left(\frac{M}{2R_b N_0/2}\right)\right]$$

(4.31)

The average power requirement in M-PPM reduces with the increase of M with respect to OOK. In M-PPM as M increases, the average power efficiency improves while the bandwidth efficiency is reduced. This is because of high peak-to-average ratio and increased signal-set dimensionality. There is no threshold detection required at the receiver for PPM unlike the OOK detection. However, symbol and chip synchronization requirements at the receiver make it necessary to have two clocks one at, a symbol rate and another at slot. Even though, PPM offers higher average power efficiency but due to its poor bandwidth efficiency it is more susceptible to multi-path-induced ISI as compared to NRZ OOK.

In the absence of multi-path distortion, an optimum ML receiver for M-PPM employs a continuous-time filter matched to one chip when output is sampled at the chip rate. Each block of M samples is passed to a block decoder which makes a symbol decision yielding $\log_2 M$ information bits. In the presence of multi-path, MLSD, equalization, and trellis-coded modulation [11-12] are used.

4.6.3   M-Level Pulse Amplitude Modulation

We have till now discussed the two level (ON and OFF) digital schemes. These are power efficient, but as the maximum symbol length increases, they require more bandwidth. Multi-level modulation schemes can achieve better bandwidth efficiency at the cost of higher power requirement. Pulse Amplitude Modulation (M-PAM) is one of the multi-level schemes used in optical wireless systems. In the case of M-level PAM the discretized amplitude levels of the optical pulse are modulated according to the amplitude of the analog signal. For M-PAM, M represents the amplitude levels of the optical signal, i.e., one of M possible amplitude levels is transmitted by the transmitter to represent a specific value. Instead of two levels, as in the case of OOK, there are M possible symbol levels with symbol period of T_b. The time varying intensity signals of M-PAM are represented in their basis functions as:

$$x(t) = \sum_{k=-\infty}^{\infty} \frac{2P}{M-1}\sqrt{T_b}[a(k)-1]\psi_{M-PAM}(t-kT_b) \tag{4.32}$$

with $a(k)$ uniformly distributed over M amplitude levels. The basis function ψ_{M-PAM} is same as that of OOK, and therefore, the non-negative condition of $x(t)$ is satisfied. For the power constraint the average amplitude of $x(t)$ is limited to P.

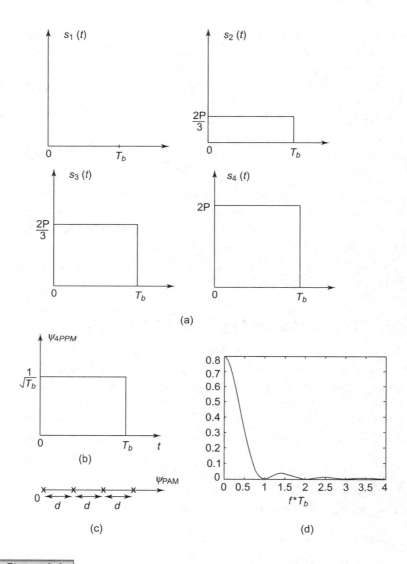

Figure 4.4 (a) 4-PAM Signals Set (b) Basis Function (c) Constellation with $d = 2P\sqrt{(T_b)/(M-1)}$ (d) 4-PAM PSD for $P = 1$ and $T_b = 1$.

The bandwidth of the PAM signal is obtained from the continuous part of power spectrum given by [13] as:

$$S_{M-PAM}(f) \approx P^2 T_b \frac{M+1}{3(M-1)} \mathrm{sin}\, c^2(\pi f T_b) \tag{4.33}$$

The PSD of 4-PAM signal is plotted in Fig 4.4 (d). The first-null bandwidth is same as for OOK; $BW_{M\text{-}PAM} = R_b$, but the bandwidth requirement of M-PAM reduces as more than one bit is sent over one symbol now. In comparison with OOK the bandwidth is:

$$BW_{M-PAM} = \frac{R_b}{\log_2 M} = \frac{1}{\log_2 M} BW_{OOK} \qquad (4.34)$$

The bandwidth efficiency of PAM can, therefore, be expressed as:

$$\frac{R_b}{BW_{M-PAM}} = \log_2 M \ \text{bits/s/Hz} \qquad (4.35)$$

In signal space, all the M-PAM signals are equally spaced in one-dimension (as they all are expressed by the same basis function) as shown in Fig. 4.4. For equal probable transmission of each symbol the symbol error is expressed as [14]:

$$P_e(sym) = 2\frac{M-1}{M}Q\left(\frac{P_{M-PAM}}{M-1}\sqrt{\frac{\log_2 M}{R_b N_0/2}}\right) \qquad (4.36)$$

The BER will depend on the coding. For Gray coding the BER value can be obtained as:

$$P_e(b) \approx 2\frac{M-1}{M\log_2 M}Q\left\{\frac{P_{M-PAM}}{M-1}sqrt\left(\frac{\log_2 M}{R_b N_0/2}\right)\right\} \qquad (4.37)$$

By inverting (4.37) the power requirement of M-PAM scheme is:

$$P_{M-PAM} = (M-1)\sqrt{\frac{R_b N_0/2}{\log_2 M}}Q^{-1}\left(\frac{1}{2}P_e(b)\frac{M}{M-1}\log_2 M\right) \qquad (4.38)$$

4.6.4   Optical Sub-Carrier Modulation

In the optical wireless sub-carrier modulation systems, the base-band signal modulates the sinusoidal electrical RF sub-carrier, typically ranging from 10 MHz to 10 GHz, either in any of the analog modulation schemes, i.e., AM, FM or PM, or of the digital modulation schemes, i.e., OOK, PPM, BPSK, QAM, etc., which subsequently intensity-modulates the optical carrier. When a bit stream modulates a single radio frequency, which is then further used to modulate optical carrier, the modulation is called the single-sub-carrier modulation (SSM). However, when a group of bit streams modulates different radio frequencies which are frequency multiplexed, the modulation is known as multiple-sub-carrier modulation (MSM) [15]. At the receiver end, the optical signal is detected like any other signal and the rest of the processing to separate the sub-carriers and extract the data from each sub-carrier is done electrically. Because the sub-carrier is typically a sinusoidal signal with positive and negative amplitude, a dc bias is added to satisfy the requirement of the transmitted optical signal to be non-negative. The sub-carrier systems have the advantage of being able to use microwave devices for modulation/detection purpose. These have better availability and the technology is also more advanced as compared to the optical components. Further, advanced modulation formats can be used for RF sub-carrier to take the advantage of their detection gain.

Figure 4.5 (a) shows a QPSK signal waveform with the sub-carrier frequency equal to $1/T_b$. The base-band signal digitally modulates the phase of the RF sub-carrier which in turn intensity-modulates the laser intensity signal [16]. The orthogonal basis functions for the QPSK signal, shown in Fig. 4.5b, are expressed as:

$$\psi_1(t) = \sqrt{\frac{2}{T_b}} \left[\cos\left(\frac{2\pi}{T_b} nt \right) \right] \tag{4.39}$$

and

$$\psi_2(t) = \sqrt{\frac{2}{T_b}} \left[\sin\left(\frac{2\pi}{T_b} nt \right) \right] \tag{4.40}$$

for some integer $n \geq 1$, $f_c = \dfrac{n}{T_b}$ is the sub-carrier frequency. Since the basis functions take on negative amplitudes at intervals within the symbol period, a DC bias (=P) must be added to ensure the non-negativity constraint is met. The transmitted optical intensity modulated signal is expressed as:

$$x_{sub-QPSK}(t) = \sum_{k=-\infty}^{\infty} \left\{ \begin{array}{l} \dfrac{\sqrt{T_b}P}{2}\left[\cos\left(\dfrac{\pi a_i[k]}{2} \right)\psi_1(t-kT_b) + \sin\left(\dfrac{\pi a_i[k]}{2} \right)\psi_2(t-kT_b) \right] \\ \\ \qquad\qquad\qquad\qquad\qquad\qquad + Prect\left(\dfrac{t-kT_b}{T_b} \right) \end{array} \right\}$$

$$\tag{4.41}$$

where $a_i(k) = 0, 1, 2$ and 3 and the maximum amplitude of the data is fixed as $P\sqrt{T_b}/2$.

The constellation of sub-carrier QPSK will be same as base-band QPSK as shown in Fig. 4.5 and the minimum Euclidean distance between the points is:

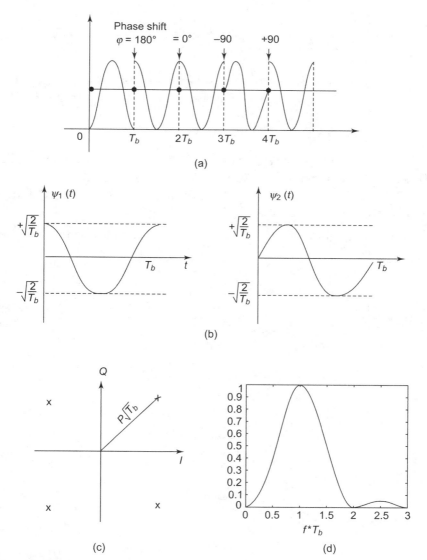

(a)

(b)

(c)

(d)

Figure 4.5 (a) QPSK Sub-carrier Signal Set with Sub-carrier Frequency = $1/T_b$, P = Average Power Transmitted (b) Basis Function (c) Constellation Diagram (d) PSD of the Continuous Part of the Spectrum.

$$d_{min} = P\sqrt{2T_b} \tag{4.42}$$

In the case of QPSK modulation, the symbol error probability is:

$$P_e(sym) \approx 2Q\left(P\frac{1}{\sqrt{R_b N_0/2}}\right) \tag{4.43}$$

The first null bandwidth of the sub-carrier PSK modulation can be calculated from its continuous spectral density expression given below:

$$S_{QPSK}(f) = P^2 T_b[\sin c^2 \pi(f + f_c)T_b + \sin c^2 \pi(f - f_c)T_b] \tag{4.44}$$

The normalized spectrum is same as QPSK signal as shown in Fig. 4.5 (d). There is less signal energy at low frequencies as QPSK is a passband modulation scheme. The bandwidth of the signal is, $BW_{QPSK} = R_b$ and the bandwidth efficiency obtained as 1.

The block diagram for the sub-carrier modulation is given in Fig. 4.6.

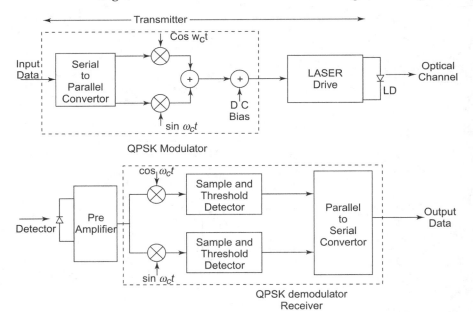

Figure 4.6   **Block Diagram for Optical SCM Scheme with QPSK Modulation.**

Pulse quadrature amplitude modulation (QAM) is yet another popular sub-carrier modulation schemes used in wireless optical channel; hence needs a mention in some details. M^2-QAM consists of an in-phase and a quadrature component basis functions, which are orthogonal to each other with M amplitude levels, as shown in Fig. 4.7. The in-phase and quadrature components basis functions, $\psi_I(t)$ and $\psi_Q(t)$, as in the case of QPSK, can be expressed as:

$$\psi_I(t) = \sqrt{\frac{2}{T_b}} \cos\left(\frac{2\pi}{T_b} nt\right) \tag{4.45}$$

$$\psi_Q(t) = \sqrt{\frac{2}{T_b}} \sin\left(\frac{2\pi}{T_b} nt\right) \tag{4.46}$$

DC is added to the transmitted signal for the non-negative condition of the signal. As in (4.41), the transmitted signal for M^2-QAM can thus expressed as:

$$x_{sub-QAM}(t) = \sum_{k=-\infty}^{\infty} \frac{\sqrt{T_b}P}{2(M-1)}\{a[k]\psi_I(t-kT_b) + b[k]\psi_Q(t-kT_b)\} + Prect\left(\frac{t-kT_b}{T_b}\right) \tag{4.47}$$

for $a[k], b[k] \in \{-(M-1), -(M-3), \dots (M-1)\}$. There will be M^2 symbols for M amplitude values because independent data modulates the two basis function at each symbol instant.

The constellation of sub-carrier M^2-QAM is a two-dimensional regular array of points. The Euclidean distance between the points is depended on the amount of DC bias added and can be obtained as:

$$d_{min} = \frac{P}{M-1}\sqrt{\frac{2\log_2 M}{R_b}} \tag{4.48}$$

From union bound approximation in (4.13) the symbol error probability is approximated by assuming d_{min} as the average minimum distance between the different points in the signal space. An estimate for the symbol error is then formed as:

$$P_e(sym) \approx \frac{4(M-1)}{M}.Q\left(\frac{P}{M-1}\sqrt{\frac{\log_2 M}{2R_bN_0/2}}\right) \tag{4.49}$$

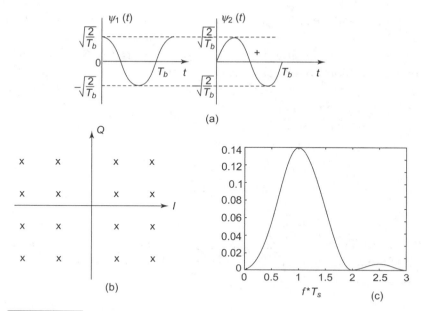

Sub-carrier QAM (a) Basis Functions (b) 16-QAM Constellation (c) PSD for 16-QAM with *P* = 1 and *T_b* = 1.

With Gray coding approximation, the probability of bit error for M^2-QAM is:

$$P_e(b) \approx \frac{2(M-1)}{M \log_2 M} \cdot Q\left(\frac{P}{M-1} \sqrt{\frac{\log_2 M}{2R_b N_0 / 2}} \right) \qquad (4.50)$$

The continuous power spectral density of M-QAM signal can be obtained as:

$$S_{QAM}(\omega) = P^2 T_b \frac{M+1}{12(M-1)} \left[\sin c^2 \left(\frac{\omega T_b}{2} - \pi \right) + \sin c^2 \left(\frac{\omega T_b}{2} + \pi \right) \right] \qquad (4.51)$$

Figure 4.7(c) gives the spectrum of the continuous portion of the PSD for 4-QAM. Similar to QPSK, QAM also has limited energy in the low frequency region. The first-null bandwidth and the bandwidth efficiency of M-QAM are respectively given as:

$$BW = \frac{2R_b}{\log_2 M^2} \quad \text{and} \quad \frac{R_b}{BW} = \log_2 M \text{ bits/s/Hz} \qquad (4.52)$$

As an example of a FSO MSM system [17], we consider a QAM-MSM system. The block diagram of the QAM-MSM system is given in Fig. 4.8. The binary input data stream is separated into $2N$ separate data streams, which is then transmitted using N sub-carrier RF frequencies. For each data stream of bit rate $1/T_b$ and a symbol set of size L_n, the average bit rate is $R_b = \dfrac{1}{T_b} \sum\limits_{n=1}^{N} \log_2(L_n)$. If $x_{n,k}$ and $y_{n,k}$ are the orthogonal symbols transmitted on data stream n on the in-phase and quadrature component at a delay kT_b and with a power coefficient P_n, the sub-carrier MSM modulated sinusoidal signal can be written as:

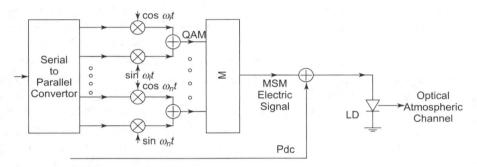

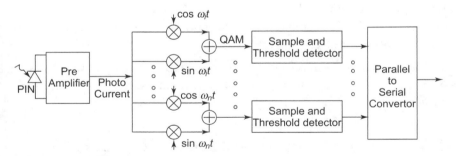

Figure 4.8 **Block Diagram of the FSO MSM System.**

$$x_{MSM-QAM}(t) = \sum_{n=1}^{N} \sum_{k=-\infty}^{\infty} \sqrt{T_b P_n /2} [x_{n,k}\psi_{In}(t - kT_b) + y_{n,k}\psi_{Qn}(t - kT_b)]$$

$$+ DC(T_b, P_n, \psi_{In}, \psi_{Qn})$$

$$(4.53)$$

where, ψ_{In} and ψ_{Qn} are the in-phase and quadrature basis functions of the modulated sub-carrier indicating its pulse shape defined in (4.45) and (4.46), respectively. DC bias is added to ensure the non-negativity constraint is met. For the case of single carrier QAM, from (4.53) the transmitted signal will reduce to (4.47) for $M = 2$.

After optical-to-electrical conversion at the receiver, the sub-carrier can be demodulated and detected using a standard QAM receiver.

In MSM, the symbol rate of each sub-carrier is reduced relative to that of a SSM with the same total bit-rate. Hence each sub-carrier becomes a narrow band signal, and therefore, experiences very less distortion. High speed OOK or PPM, which are wide-band signals and suffer from ISI due to multi-path dispersion above 100 Mbs can be used in MSM system. At the receiver, the individual bit steams are recovered using multiple bandpass demodulators. On the wrong side, the optical power efficiency of MSM schemes is much lower than for OOK or PPM modulation [15], an N-carrier transmission requires $5\log_{10}N$ db more optical power than the corresponding single sub-carrier scheme. This is to insure that $x(t)$ is positive and the amplitude of each sub-carrier does not exceed P/N. Though SSM and MSM are less power efficient than OOK or PPM because of the DC power required, but advantages gained with MSM are important. MSM is more suitable for transmission of multiplexed bit streams; hence high data rates are achievable without the need of adaptive equalization to eliminate ISI as in the case of OOK or PPM. In addition, both SSM and MSM schemes can provide higher immunity than OOK to the low frequency noise.

In the following sections, few more modulation schemes, which have been proposed for possible application in optical wireless, are discussed in brief.

4.6.5   Differential Pulse Position and Digital Pulse Interval Modulation

Differential PPM (DPPM) [18] is a modification of PPM with higher transmission capacity compared to PPM. It achieves the higher capacity by eliminating all the unused time slots or 'chips' within each symbol. Each symbol is initiated with a pulse eliminating the need of any symbol synchronization.

A DPPM symbol is obtained from the corresponding PPM symbol by deleting all of the redundant 'OFF' chips following the 'ON' chip. For a fixed M, M-DPPM will have a higher duty cycle and is therefore less power efficient than M-PPM. However, for a fixed average bit-rate and bandwidth, DPPM can employ higher M resulting in a net increase in average power efficiency. The symbol set using rectangular pulses for DPPM can be represented as;

$$x_n(t) = \begin{cases} 0.....................t < nT_c \\ P.........nT_c \leq t \leq (n+1)T_c \end{cases} \tag{4.54}$$

where, T_c is the chip period. The 4-DPPM signal sets are shown in Fig. 4.9. For comparison the 4-bit OOK symbol and 4-PPM signal sets are also shown in the figure. We can observe that in DPPM all the OFF chips are omitted after the ON chip. When compared to 4-PPM symbol each symbol consists of four chips, of which one is *high* and three are *low*, while .the symbol length in DPPM is non-uniform. The symbol boundaries are not known prior to detection. Hence, optimal soft decoding of DPPM requires the use of MLSD, even in the absence of coding or ISI. If hard decoding is used, DPPM is easier to decode than PPM, since the former requires no symbol-level timing recovery.

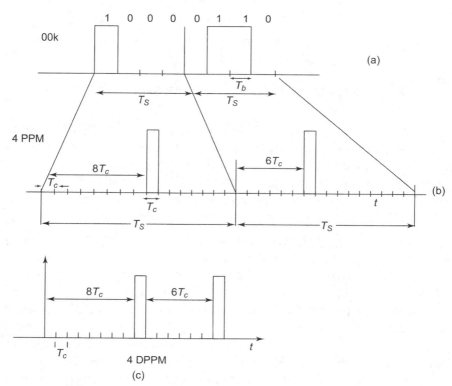

Figure 4.9 4-DPPM for 4-bit Symbols: 1000 and 0110.

Digital pulse interval modulation (DPIM) [19] scheme is another potential alternative to PPM. Data is encoded in a number of discrete time slots or 'chips' as in PPM. But unlike PPM all the unused time slots are eliminated from within each symbol more like the DPPM. The difference from DPPM is that in order to avoid symbols in which the time between adjacent pulses is zero, an additional guard slot is added to each symbol immediately following the pulse. Thus, a symbol which encodes L bits of data is represented by a pulse of constant power in one slot followed by k slots of zero power, where, $1 \leq k \leq M$ and $M = 2^L$. DPIM requires no symbol synchronization since each symbol is initiated with a pulse. The minimum and maximum symbol lengths are $2T_c$ and $(M + 1)T_c$ seconds, respectively. The DPIM waveforms for the same PPM symbols of Fig. 4.9 are shown in Fig. 4.10.

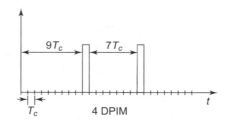

Figure 4.10   **DPIM Signal Waveform for 4-bit Symbols 1000 and 0110.**

Besides the above discussed digital schemes, there are other more complex schemes proposed. *Hybrid Pulse Amplitude and Position Modulation (M-n-PAPM)* [20] has the information represented both by the amplitude and by the position of the pulse. The hybrid PAPM is a very useful variation, which is employed in systems that require adaptation between the high spectral and power efficiency requirements.

Differential Phase-Shift Keying (DPSK) has received significant interest from the FSO community in recent times since it is power-efficient and has a 3 dB sensitivity improvement over OOK with an improved spectral efficiency. Binary information is conveyed with two orthogonal symbols represented by the relative phase between two differentially encoded bits: '0' by no-phase change and a '1' by a π phase difference or visa-versa [21]. Unlike OOK, no adaptive threshold is required in DPSK to improve BER. These benefits of DPSK over OOK comes at the cost of increased complexity, requiring a phase modulator and differential pre-coding in the transmitter, and an optical delay-line interferometer demodulator and balanced detection in the receiver.

For the sake of comparision, plot of bandwidth efficiency and power gain is given in Fig. 4.11 for the different modulation schemes discussed above.

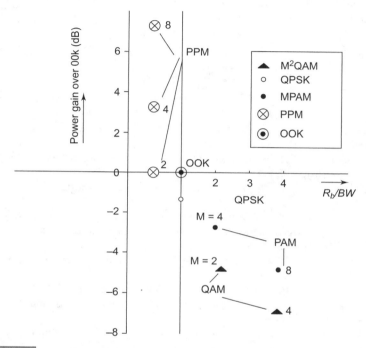

Figure 4.11 Power Gain and Bandwidth Efficiency of Different Schemes.

Summary

In this chapter, several modulation schemes used in wireless optical systems have been discussed. The wireless optical intensity channel imposes the constraint of the signal to be non-negative. The transceiver structure should be simple; therefore, intensity modulation and direct detection (IM-DD) transceivers are commonly used. Also, the average power transmitted is constraint due to eye safety consideration. The selection criterion for the modulation scheme, like in any other communication system, is power and bandwidth efficiency. While the power efficiency limits the link length, bandwidth efficiency restraints the achievable bit rate of the link.

The most popular modulation format for FSO links are the binary level OOK and M-PPM. Multi-level format M-PAM is proposed to provide high spectral efficiency at the expense of higher BER. Sub-carrier MSM system, though provide reduced ISI by multiplexing lower rate signals on multiple sub-carriers, but at the cost of higher input power.

References

1. John M. Senior, *Optical Fiber Communications, Principles and Practice;* Prentice-Hall International, 1985.
2. John G. Proakis and Masoud Salehi, *Fundamentals of Communication Systems,* Pearson Education, 2008.
3. Haykin S., *Digital Communications,* John Wiley & Sons, New York, NY, 1988.
4. Proakis J.G., *Digital Communications,* McGraw-Hill, New York, NY, 1983.
5. Lee E.A. and Messerschmitt, D.G. *Digital Communication,* Kluwer Academic Publishers, Boston, MA, 2nd edition, 1994.
6. Kahn J.M. and Barry J.R., 'Wireless Infrared Communications,' *Proc. IEEE,* Feb. 1997, pp. 265–98.
7. Park H. and Barry J.R., 'Modulation analysis for wireless infrared communications' *Proc. IEEE International Conference on Communications,* pp 1182–1186, 1995.
8. *Infrared Data Association: serial infrared physical layer specification,* Version 1.3, www.irda.org, 1998.
9. *Infrared Data Association serial infrared physical layer specification,* Version 1.4, www.irda.org, 2001.
10. Hranilovic S. *'Wireless Optical Communication Systems',* Springer Science, 2005.
11. Lee D.C., Kahn J.M. and Audeh M.D., 'Trellis-Coded Pulse-Position Modulation for Indoor Wireless Infrared Communications', *IEEE Trans. Commun,* Vol. 45, no. 9, Sept. 1997, pp 1080–1087.
12. Lee C.M. and Kahn J.M. 'Coding and equalization for PPM on wireless infrared channels', *IEEE Trans on Comm,* 47(2): 255–260, February 1999.
13. Bruce Carlson A., *Communication Systems: An Introduction to Signals and Noise in Electrical Communication,* 3rd ed., McGraw-Hill, New York, 1986.
14. Barry J. R., Lee E. A. and Messerschmidt D. G., *Digital Communication,* Kluwer Academic Publishers, 2004.
15. Carruthers J.B. and Kahn J.M., 'Multiple-sub-carrier modulation for nondirected wireless Infrared communication', *IEEE Journal on Selected Areas in Comm,* Vol. 14, No. 3, April 1996, pp. 538–546.
16. Li J., John Q. Liu and Desmond P. Taylor, 'Optical Communication using Sub-carrier PSK Intensity Modulation through Atmospheric Turbulence Channels', *IEEE Trans on Comm,* Vol. 55, no. 8, Aug. 2007, pp 1598–1606.

17. Ohtsuki T., 'Multiple Sub-carrier modulation in optical wireless communication', *IEEE Comm. Magazine*, March 2003, Vol. 41, no. 3, pp. 74–79.

18. Shiu D. and Kahn J.M., 'Differential pulse position modulation for power-efficient optical communication,' *IEEE Trans. on Comm.*, Vol. 47, no. 8, Aug. 1999, pp 1201–1210.

19. Ghassemlooy Z., and Hayes A.R., 'Pulse interval modulation for IR communications,' *Int. J. Commun. Syst. – Special Issue*, 13, 2000, pp. 519–536.

20. Zeng Y., Roger J. Green, Sun A. and Leeson Mark S., 'Tunable pulse amplitude and position modulation technique for reliable optical wireless communication channels,' *Journal of Comm*, Vol. 2, No.2 march 2007, pp 22–28.

21. Le Ngugen Binh, *'Digital Optical Communications'*, CRC Press, Taylor & Francis Group, 2009.

Further Reading

1. Barry J.R., *Wireless Infrared Communications*, Kluwer, 1994.

2. Ghassemlooy Z. et al., 'Digital pulse interval and width modulation for optical fibre communication,' Proc. SPIE, Vol. 2614, 1995, pp. 60–68.

3. Davidson F.M. and Sun X., "Slot Clock Recovery in Optical PPM Communication Systems with Avalanche Photodiode Photo-detectors," IEEE Trans. Commun., Vol. 37, no. 11, Nov. 1989, pp. 1164–71.

4. Moreira A.J.C. et al., 'Modulation methods for wireless infrared transmission systems—performance under ambient light noise and interference,' *SPIE*, Vol. 2601, 1995, pp. 226–237.

5. Audeh M.D., Kahn J.M., and Barry J.R., 'Performance of pulse position modulation on measured nondirected indoor infrared channels,' *IEEE Trans. Commun.*, Vol. 44, pp. 654–659, June 1996.

6. Sato M., Murata M. and Namekawa T., 'A new optical communication system using the pulse interval and width modulation code', *IEEE Trans on Cable Television*, Vol. CATV-4, no. 1, pp. 1–9, 1979.

7. Huang W., Takayanagi J., Sakanaka T., Nakagawa M., 'Atmospheric optical communication system using sub-carrier PSK modulation', *IEICE Trans Comm*, Vol. E76-B, pp.1169–1176, May 1993.

8. Wakafuji K. and Ohtsuki T., 'Performance Analysis of atmospheric optical sub-carrier multiplexing systems and CDM systems', *Journal of Lightwave Tech*, Vol. 23, No. 4, April 2005. pp. 1676–1682.

9. Shiu D. and Kahn J.M. 'Shaping and non-equiprobable signalling for intensity-modulated signals', *IEEE Trans on Information Theory*, 45(7): 2661–2668, Nov 1999.

10. Lee G.M. and Schroeder G.W. 'Optical pulse position modulation with multiple positions per pulsewidth', *IEEE Trans on Comm*, 25: 360–364, March 1977.

11. Ghassemlooy Z., Hayes A.R., Seed N.L., and Kaluarachchi E.D. 'Digital pulse interval modulation for optical communications,' *IEEE Comm. Magazine*, 36(12): 95–99, December 1998.

12. Audeh M.D., Kahn J.M., and Barry J.R. 'Decision-feedback equalization of pulseposition modulation on measured nondirected indoor infrared channels', *IEEE Trans Comm*. 47(4): 500–503, Feb 1999.

13. Hranilovic S. and Johns D.A. 'A multilevel modulation scheme for high-speed wireless infrared communications', Proceedings of the *IEEE International Symposium on Circuits and Systems*, Vol. VI, pages 338–341, 1999.

14. Garcia-Zambrana A. and Puerta-Notario A. 'Improving PPM schemes in wireless infrared links at high bit rates', *IEEE Communications Letters*, 5(3): 95–97, March 2001.

15. You R. and Kahn J.M. 'Average power reduction techniques for multiple sub-carrier intensity-modulated optical signals', *IEEE Transactions on Comm*. ,49(12): 2164–2171, December 2001.

16. Forney G.D. Jr. and Ungerboeck G., 'Modulation and coding for linear Gaussian channels', *IEEE Trans on Information Theory*, 44(6):2384–2415, October 1998.

Diversity and Detection Techniques in Optical Fading Channel

Chapter 5

Optical wireless channel behaves as a fading channel due to the turbulence and multipath phenomenon in the atmosphere. There is a significant probability that during transmission the channel may be in deep fade and hence can cause significant performance degradation with very poor BER. The error probability in the fading channel is much higher than in a non-fading the AWGN channel. This motivates us to investigate various techniques, which can be used to improve the performance. Like in RF channels, the performance in the optical wireless fading channels can be improved by *diversity techniques* or by *coding*. The diversity techniques operate over time, frequency or space. The basic idea behind any diversity technique is to send multiple copies of the signal at different time or frequency or on different paths. These multiple independently faded replicas of data symbols so obtained at the receiver end are processed appropriately, and then a more reliable detection can be achieved. There are many sophisticated schemes which exploit not only the channel diversity, but at the same time efficiently use the degrees of freedom in the channel. These schemes can provide array and multiplexing gains in addition to diversity gains. The system performance and robustness can also be improved by coding the signal before transmission. Many coding schemes used in the FSO system will be discussed in Chapter 7.

In Section 5.1 of this chapter, we look at the various issues of detection over the optical fading channels. In the previous chapter, in order to know the characteristics of modulation schemes, the channel was chosen to be time invariant linear filter with AWGN. In practical systems, the channel characteristics are time varying, band limited, and therefore, more complex models as discussed in Chapter 3 are required to describe their behavior. In Section 5.2, we discuss the signal models used for optical detection in FSO. Section 5.3 deals with the detection over the Single Input Single Output (SISO) link over the fading channel. In the subsequent Sections, 5.4 to 5.7, different diversity schemes and their detection are described.

The derivations in this chapter inadvertently make use of a few key results in vector detection under AWGN as was done in Chapter 4. The detection of the transmitted signal is done at the electrical level, therefore, as in Chapter 4, we use the baseband electrical channel model with optical channel constraints of non-negative signal amplitude and average power constraints.

5.1 DETECTION IN AN OPTICAL FADING CHANNEL

For better understanding of the detection in optical fading channel, we start with a simple detection problem. For simplicity, let us assume a flat fading channel model. The received baseband signal can be expressed as:

$$y(t) = h(t) * x(t) + w(t) \tag{5.1}$$

where, $x(t)$ is the input optical signal, $y(t)$ is the detector current and AWG noise $w(t)$ is a Gaussian random variable with zero mean and $N_0/2$ variance. We can assume light intensity fading to be either log-normal or exponential. In this case, the channel characteristics $h(t)$ may be treated as constant over the signal duration T_s, which is a slow fading, frequency non-selective channel. In order to keep the mathematics simple, we look at 2-PPM modulation scheme with two slots in the symbol period of T_s.

The input pair of optical intensity samples at two times instant as given in Fig. 5.1 are expressed as:

$$x_A = \begin{pmatrix} x[0] \\ x[1] \end{pmatrix} = \begin{pmatrix} 2P \\ 0 \end{pmatrix} \tag{5.2}$$

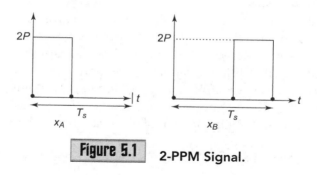

Figure 5.1   2-PPM Signal.

$$x_B = \begin{pmatrix} x[0] \\ x[1] \end{pmatrix} = \begin{pmatrix} 0 \\ 2P \end{pmatrix} \tag{5.3}$$

The detection of the transmitted signal is based on detector current by:

$$y = \begin{pmatrix} y[0] \\ y[1] \end{pmatrix} \tag{5.4}$$

By the ML detection rule the log-likelihood ratio (LLR) $\Lambda(y)$, in this case is derived as [1]:

$$\Lambda(y) \underset{x_B}{\overset{x_A}{\gtrless}} 0 \tag{5.5}$$

where, $\Lambda(y) = \ln\left\{\dfrac{f(y/x_A)}{f(y/x_B)}\right\}$.

Assuming the detector current to be Gaussian in nature, it can be seen that for the flat fading, frequency non-selective channel, if x_A is transmitted then $y[0]$ is a Gaussian random vector with zero mean and variance of $(P^2T_s + N_0/2)$ and $y[1]$ will have a mean of zero and variance of $N_0/2$. Similarly, if x_B is transmitted then $y[0]$ Gaussian random vector has a variance of $N_0/2$, and $y[1]$ has a variance of $(P^2T_s + N_0/2)$. Both, $y[0]$ and $y[1]$ are independent. Hence, the LLR is calculated to be:

$$\Lambda(y) = \frac{\left[\,|y[0]|^2 - |y[1]|^2\,\right]P^2T_s}{(P^2T_s + N_0/2)N_0/2} \tag{5.6}$$

Therefore, by ML detection rule, x_A is decided to be transmitted if $|y[0]|^2 > |y[1]|^2$, and is decided as x_B otherwise. In other words, the detector projects the received vector y onto each of the two possible transmits vectors x_A and x_B, and compares the energies of the projections and, therefore, the detector can be considered to be a square-law detector. From above, as the received signals $|y[0]|$ and $|y[1]|$ are independent Gaussian random variables, hence, in (5.6) $|y[0]|^2$ and $|y[1]|^2$ are exponentially distributed with mean $(P^2 T_s + N_0/2)$ and $N_0/2$, respectively, if x_A is assumed to be transmitted. The probability of symbol error can then be computed as:

$$P_e(sym) = P\left[|y[1]|^2 > |y[0]|^2 |x_A\right] = \frac{1}{1 + \dfrac{P^2 T_s}{N_0/2}} \tag{5.7}$$

The electrical SNR after detection, which is the ratio of average received signal energy per symbol time to the noise energy per symbol time, is given in this case of 2-PPM as:

$$SNR = \frac{P^2 T_s}{N_0/2} \tag{5.8}$$

Writing (5.7) in terms of SNR, the symbol error rate (SER) can be expressed as:

$$P_e(sym) = SER = \frac{1}{(1 + SNR)} \tag{5.9}$$

or, the error increases linearly with decreasing SNR.

We next compare this performance with detection in an AWGN channel without fading, where the output of the channel is expressed as:

$$y(t) = x(t) + w(t) \tag{5.10}$$

The error probability calculated in this case from Chapter 4 can be expressed in terms of SNR as;

$$SER = Q\sqrt{SNR} \tag{5.11}$$

where, $Q(\zeta)$ is the complementary cumulative distribution function which decays exponentially with ζ^2 as discussed in Chapter 4. It is repeated here for convenience:

$$Q(\zeta) < e^{-\zeta^2/2} \qquad\qquad \zeta > 0$$

$$Q(\zeta) > \frac{1}{\sqrt{2\pi}\zeta}\left(1 - \frac{1}{\zeta^2}\right)e^{-\zeta^2/2} \quad \zeta > 1$$

Therefore, in the case of AWGN channel the SER decreases exponentially with decreasing SNR. For example, to obtain a SER = 10^{-6} in the case of fading channel from (5.9) one would require SNR of 60 dB, which is very high to be obtained practically. On the other hand to obtain the same error probability of 10^{-6} in the case of AWGN channel a SNR of just about 9.5 dB is needed, which is a reasonable figure. Certain improvement though in case of random channel can be obtained if the receiver knows the channel gain even though they are still random.

The main reason of the poor detection performance in the fading channel is due to the fact that the channel gain is random, and there is a significant probability that the channel is in a *deep fade* when the signal is being transmitted as shown in Fig.5.2.

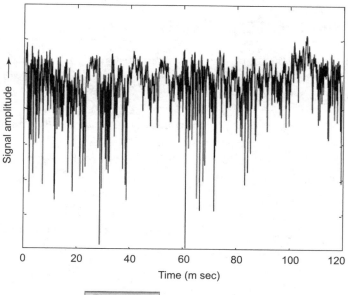

Figure 5.2   Fading Signal.

Without detailed mathematical derivations, in the case of fading channel, similar to (5.11), one can express the SER with complete statistical channel information at the receiver as [2]:

$$SER = E\left[Q\sqrt{|h|^2 SNR}\right] \tag{5.12}$$

The quantity $|h|^2 SNR$ is now the instantaneous received SNR. Due to atmospheric conditions during deep fade instants even at high SNR values, the quantity $|h|^2$ *SNR* can become $\ll 1$, leading to a significant increase in SER; the signal level is now reduced considerably and is of the same order as that of the noise and hence the error probability becomes significant. In contrast, in AWGN channel, the only possible error mechanism is due to the additive noise. Thus, the error probability performance in the AWGN channel is much better and can be improved by increasing the signal strength, while in case of fading channel increasing signal strength cannot help when the signal is in deep fade. We conclude that in atmosphere channel, high-SNR error events most often occur because the channel is in deep fade and not as a result of the additive noise being large. Therefore, alternative measures such as diversity and/or coding are resorted to for the improvement in the system performance.

5.2   SIGNAL MODELS FOR DETECTION

In free space optics, there are two models, which can be used for the detection of the received signal depending on the signal strength and the noise present.

Gaussian model   In the case of high incident signal power, the detected signal in the FSO system can be modeled as Gaussian distributed as in the case of its RF counterpart. The block diagram of the receiver structure is shown in Fig. 5.3.

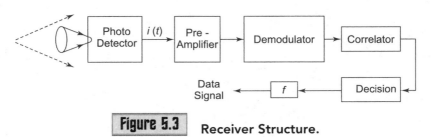

Figure 5.3   Receiver Structure.

On the incident of the optical beam, the photodetector converts the incident modulated light to photocurrent. The receiver integrates the received photocurrent for an interval $T_0 \leq T_b$ during each bit interval T_b. At the end of the integration interval, the resulting detected average electrical current including the thermal noise current can be expressed as [3]:

$$I_{ph} = \Re(P_s + P_b) + I_{th} \tag{5.13}$$

where, P_s is the received signal light power and P_b is the ambient light intensity assumed to be constant during the integration time T_0 and $\mathfrak{R}$ is the responsitivity of the detector in A/W. The thermal noise current I_{th} is Gaussian with zero mean and is independent of whether the received bit is *On* or *Off*. The detected signal or the current I_{ph} can be approximated to be Gaussian in this case. In most of the practical IM/DD free-space optical communication systems, the receiver signal-to-noise ratio is limited by shot noise caused by ambient or background light which is much stronger than the desired signal, and/or by the thermal noise in the amplifier electronics following the photo detection. Both, the ambient and thermal noise are usually modeled with high accuracy as AWGN which is statistically independent of the desired signal. Thus, the Gaussian model of an FSO system is adopted in this case when high signal energies are considered and where the shot-noise in the information bearing component is negligible relative to the shot-noise inherent to the background radiation.

Poisson Photon-counting model   In the high performance system and in general at low SNR, the signal-dependent shot-noise becomes the limiting factor. In such cases photon-counting model is more appropriate and the photo detector analysis is based on general photon-counting model of photo detection, where the incident photons produce photoelectrons as a Poisson point process, which contribute to the detector current at the output [4]. The rate of the Poisson process is proportional to the total power incident on the photo detector. When the electrical current is integrated over each bit interval it then produces a set of statistics suitable for detection. As the input to the integrator is proportional to the incident power, the resultant statistics are effectively proportional to the intensity of the incoming light constituted of the ambient light of the background and the transmitted signals which can be accurately modeled as a Poisson point process whose variance is dependent on both of these contributions. The probability mass function for the number of counts Z, in the *On* interval is Poisson distributed, and is expressed as:

$$P(Z = k) = \frac{(n_s + n_b)^k e^{-(n_s + n_b)}}{k!}, k = 0, 1, 2, \ldots \quad (5.14)$$

where, n_s is the average number of signal photoelectrons produced at the detector and n_b is the average number of photoelectrons produced per bit interval T_b due to background radiation. They can be respectively expressed as:

$$n_s = \frac{\eta P_s T_b}{hf} = \frac{\eta E_s}{hf} \qquad (5.15a)$$

and

$$n_b = \frac{\eta P_b T_b}{hf} \qquad (5.15b)$$

where η is the intrinsic efficiency of the detector and E_s is the energy in the pulse of width T_b. In the *Off* interval, the count variable Z is Poisson with parameter n_b, and also has a Poisson statistics given as:

$$P(Z = k) = \frac{(n_b)^k e^{-(n_b)}}{k!}, k = 0,1,2,\dots \qquad (5.16)$$

Let us understand the above two detection processes. When the thermal noise, which is Gaussian in nature, is high, its variance is much larger than the optical shot noise variance. In this case, the first term in (5.13) is negligible and I_{ph} can then be accurately approximated by a Gaussian random variable. Also, in the same case, when the level of background power P_b is high enough, the pdf for the count of photoelectrons (5.15 b), which is Poisson in nature, becomes increasingly Gaussian in shape [5]. When this is added to the Gaussian thermal noise the resulting pdf can also be modeled as Gaussian for the *Off* bit. On the other hand in *On* state when the incident signal power *Ps* is high, then that is also a sufficient condition for the Gaussian approximation to hold. But, in the case where the background power P_b and thermal noise power levels are both low, the Gaussian approximation ceases to yield accurate results. Therefore, at low signal levels, we must use the photodetector current model to be Poisson distributed, both, for the *On* and *Off* signal levels.

5.3 DETECTION IN SINGLE INPUT SINGLE OUTPUT TURBULENCE CHANNELS

We first describe symbol-by-symbol optimal detection scheme the for SISO link in the turbulent fading atmosphere. We have earlier discussed the ML and MAP detection in the case of non-turbulent AWGN channels in Chapter 4. Now, we derive the expression for the turbulent fading channels.

Let us consider the simple case of OOK modulation. The optical channel is assumed to have turbulence induced fluctuation of log-amplitude

X, with a marginal distribution $f_X(X)$, which is Gaussian in nature. As discussed in Chapter 3, the light intensity I is related to X as; $I = I_0 \exp(2X - E[X])$. We also assume the receiver has knowledge of the marginal distribution of the turbulence-induced fading.

At the receiver, from (5.13) the received signal intensity can be obtained after subtracting the ambient light intensity. The additive noise is Gaussian with a mean of zero and variance of $N_0/2$. The optimal MAP symbol-by-symbol detector decodes the transmitted bit $\hat{I}_t$ as [1]:

$$\hat{I}_t = \arg\max_{I_t} P(I|I_t)P(I_t) \tag{5.17}$$

where, $P(I_t)$ is the probability of the *On* or *Off* bit being transmitted, $P(I|I_t)$ is the conditional distribution that if a bit I_t is transmitted a signal level I will be received. In the case of equally probable *On* and *Off* bits, the symbol-by-symbol ML detector decodes the bit I_t as

$$\hat{I}_t = \arg\max_{I_t} P(I|I_t) \tag{5.18}$$

As discussed above, the likelihood function $\Lambda(I)$ is obtained as the ratio of the conditional probability of the transmitted bit to be *On* or *Off*; i.e., the ratio of $P(I|On)$ and $P(I|Off)$, respectively, These for the fading channel are described by [6]:

$$P(I|On, X) = \int_{-\infty}^{\infty} p(I|On, X) f_X(X) dX$$

$$= \int_{-\infty}^{\infty} \frac{1}{\sqrt{\pi N_0}} \exp\left[-\frac{\left(I - I_0 e^{2X-2E[X]}\right)^2}{N_0} \right] f_X(X) dX \tag{5.19}$$

and

$$P(I|Off) = \frac{1}{\sqrt{\pi N_0}} \exp\left(-\frac{I^2}{N_0} \right) \tag{5.20}$$

The likelihood function is then given by:

$$\Lambda(I) = \frac{P(I|On)}{P(I|Off)} = \int_{-\infty}^{\infty} f_X(X) \exp\left(-\frac{\left(I - I_0 e^{2X-2E[X]}\right)^2 - I^2}{N_0} \right) dX \tag{5.21}$$

As seen in (5.21), the likelihood ratio increases monotonically with the received signal I, therefore, ML detection can be implemented with threshold detection by setting appropriate threshold level of the received signal. With the increase of turbulence-induced fading the threshold decreases toward zero as the turbulence effects the fluctuation of the *On*-state signal level only, but does not effect the *Off*-state of signal level. Considering the case of OOK signal modulation, the bit-error rate (BER) can be calculated as;

$$BER = P(On)P(b|On) + P(Off)P(b|Off) \tag{5.22}$$

where, $P(b|On)$ and $P(b|Off)$ are the bit-error probabilities, when the transmitted bit is *On* and *Off*, respectively. At low bit rate, these can be expressed as following when there is no intersymbol interference:

$$P(b|Off) = \int_{\Lambda(I)>1} p(I|Off)\,dI \tag{5.23a}$$

$$P(b|On) = \int_{\Lambda(I)<1} p(I|On)\,dI \tag{5.23b}$$

Figure 5.4 gives the BER performance for the OOK for the case of AWGN and log-normal fading channel for different turbulent conditions.

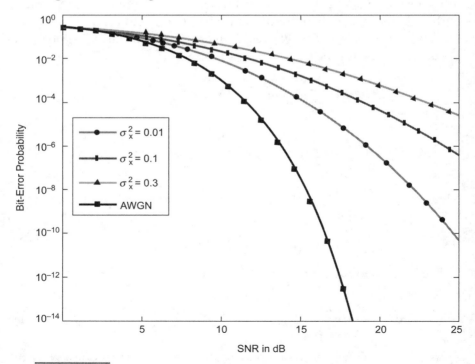

BER Performance for OOK Modulation with AWGN, and Log-normal Fading in the Atmospheric Channel.

5.4 DIVERSITY

As discussed in the above section, due to fading in the atmosphere the performances of the various modulation schemes used in the optical wireless system are poor. One finds the error probabilities for all the modulation schemes to decay very slowly: inversely proportional to SNR for high SNR values. The reason for this poor performance is that the communication depends on the strength of one single signal path and there is a significant probability that this path will be in a deep fade. When the path is in a deep fade, any communication scheme is likely to suffer from errors because the power level goes well below the detectable level as discussed above. If we ensure that the receiver receives multiple copies of these information symbols either through multiple signal paths or at different times, each of which fades independently, making sure that reliable communication is possible as long as one of the signal is strong. This technique is called diversity and it can dramatically improve the performance over fading channels.

There are many ways to obtain diversity.

➤ *Time Diversity* is applicable in a channel that has time selective fading. Diversity over time can be obtained via coding and interleaving; the information is coded and the coded symbols are dispersed over time in different coherence periods so that different parts of the codewords experience independent fades.

➤ *Frequency or wavelength Diversity* is effective when the channel fading is wavelength selective. It can be exploited by spreading the data over a frequency span larger than the coherence bandwidth of the channel.

➤ *Space Diversity:* With multiple transmit or receive antennas spaced sufficiently far apart by more than the coherence distance, then diversity can be obtained over space as well.

In optical wireless systems space and time diversity are more commonly used techniques though frequency or wavelength diversity is also possible.

5.5 SPATIAL DIVERSITY

Spatial diversity or antenna diversity can be obtained by using multiple antennas at the transmitter and/or the receiver. If the antennas at the transmitter end and/or at the receiver end are placed sufficiently far apart, more than the coherence distance in the plane of the antennas, the channel gains between different transmitter and receiver antenna pairs fade more or less independently, and hence, independent signal paths are created. The coherence distance is expressed as $(\sqrt{\lambda L})$, where L is the link length and λ is the wavelength. The required antenna separation depends on the beam divergence, local scattering environment, as well as on the optical wavelength. Typical antenna separation of the order of 10's of wavelength is sufficient to receive signals from independent fading paths. In the case of optical communication as the wavelengths are in micrometers, it is very practical to have small arrays with large number of diverse antennas and then obtain large diversity gain. This is a reasonable assumption in the case of terrestrial links. Spatial diversity reception in free-space optical communication has also been proposed and studied for both near-earth links and interplanetary links. For these links sometimes it may be difficult to satisfy this assumption in practice, in particular in very long power-limited links which often employ well-collimated beams. The antenna spacing required in such cases for uncorrelated fading may well exceed the beam diameter at the receiving end.

To keep the discussion simple, we consider the scenario when the receiver has perfect knowledge of the channel gains. This knowledge is learnt via pilot symbols. The accuracy of the channel estimate depends on the coherence time of the channel and the received power of the transmitted signal. We will look at both, the receive diversity, i.e., using multiple receive antennas (single-input-multi-output or SIMO channels) and the transmit diversity, using multiple transmit antennas (multi-input-single-output or MISO channels). Channels with multiple transmit and multiple receive antennas (MIMO) provide in addition to diversity, additional degrees of freedom for communication.

The performance improvements with the space diversity systems can be obtained due to:

➤ Antenna/Array gain: Array gain is possible through processing at the transmitter and the receiver and results in average increase in receive SNR. Transmit/receive array gain requires channel knowledge in the transmitter and receiver, respectively, and depends on the number of transmit and receive antennas.

➤ Diversity gain: Diversity gain is achieved by receiving multiple copies of the same signal through independent fading paths. This increases with turbulence in the channel, number of transmit/receive antennas and with space-time coding.

➤ Spatial multiplexing gain: Unlike the case of antenna or diversity gain, for the case of spatial multiplexed MIMO system a linear increase in capacity of channel can be achieved with no increase in transmitted power or bandwidth but with the increase in detection complexity. The capacity increase is achieved by transmitting independent data signals from the individual transmit antennas. We will discuss about the capacity of optical channel in Chapter 6.

5.5.1 Receive Diversity

In the case of receive diversity for flat fading channels with one transmit and N receiving antennas (Fig. 5.5a), the channel model for q time instant can be expressed as:

$$y_n[q] = h_n[q]x[q] + w_n[q] \text{ for } n = 1...N \tag{5.24}$$

where $w_n[q]$ is the AWG noise and is independent across the antennas. We would like to detect the transmitted signal $x[q]$, based on the received detector current at N receive antennas $y_1[q]...y_N[q]$. As the receiver antennas are spaced sufficiently far apart, we can assume that the gains $h_n[q]$ are independent and have a random distribution; hence, with these N diverse signals we can obtain a diversity gain upto N. This diversity gain is achieved due to averaging over multiple independent signal paths. This increases the probability of the overall gain.

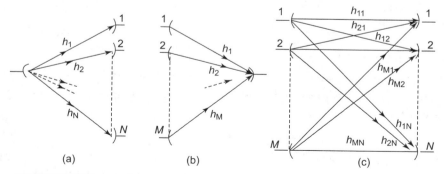

Figure 5.5 Space Diversity Schemes, (a) Receive Diversity, (b) Transmit Diversity (c) Multi-Input Multi-Output Space Diversity.

In the case of receive diversity besides the diversity gain array gain is also achieved with the increase of N. The array gain is due to the *square-law* combining of the received signals at the multiple receive antennas. Due to array gain the effective total received signal power increases linearly with N. There is a 3 dB gain obtained with every doubling of the number of antennas. On the other hand, in the case of diversity gain, unlike the array gain, because of the law of large numbers there is a diminishing marginal return with increasing N even when all the channel gain components h_n are independent.

5.5.2   Transmit Diversity

In the case of transmit diversity there are M transmit antennas and one receive antenna. This is called the MISO channel (Fig. 5.5(b)). Diversity gain of M can be obtained by two means in MISO channels. One method is by transmitting the same symbol over the M different antennas during M symbol times. At any one time, only one antenna is turned on and the rest are silent. This is a kind of repetition coding in space. It is quite wasteful of degrees of freedom. The other more useful method is by using any time diversity code of block length M. Different symbols of the code are sent over the different antennas one at a time. This provides a coding gain over the repetition code obtained in the first case. Alternatively, one can also design codes specifically for the transmit diversity system which are called the *space-time* codes. The most elegant space-time code for RF wireless system is the Alamouti scheme. The Alamouti scheme is modified for optical intensity system [7]. Optical Alamouti code is designed for 2 transmit antennas but can be generalized to more than 2 antennas to some extent.

Optical Alamouti Scheme

We describe the Alamouti scheme for 2 transmit MISO case. The flat fading channel response with two transmit and a single receive antenna is written as

$$y[q] = h_1[q]x_1[q] + h_2[q]x_2[q] \tag{5.25}$$

where, h_1 and h_2 are channel gain from transmit antenna 1 and 2 to the receiver, respectively. As in the case of conventional Alamouti scheme, in the first bit interval, antenna 1 transmits signal x_1 and antenna 2 transmits x_2, where x_1 and x_2 each range over the signal set (u_1, u_2). Considering the case of OOK modulation, the symbols can be described by:

$$u_1 = 0......0 \leq t \leq T_s,$$
$$u_2 = A,.....0 \leq t \leq T_s$$

where, A is the peak intensity of the light and T_s is the symbol duration. We observe that,

$$u_i = -u_j + A,...i \neq j \tag{5.26}$$

Therefore, we can define the complement $\overline{x}_i$ of x_i as obtained by reversing the role of *On* and *Off* bits. As an example if $x_i = u_1$, then from (5.26) $\overline{x}_i$ is;

$$\overline{x}_i \equiv u_2 = -u_1 + A$$

Next, in the second bit interval, antenna 1 transmits $\overline{x}_2$ and antenna 2 transmits x_1.

The received signals with noise, in the first and second bit intervals, respectively, therefore, can be written as:

$$y[1] = h_1 x_1[1] + h_2 x_2[1] + w_1 \tag{5.27a}$$
$$y[2] = h_1 \overline{x}_2[2] + h_2 x_1[2] + w_2 \tag{5.27b}$$

where noise w is modeled as independent zero-mean Gaussian real random variables and the channel is assumed to be constant over the symbol periods. Assuming perfect knowledge of the channel gains and the transmitted signal level at the receiver, the receiver estimates the transmitted signals. The decision metrics for the transmitted signal x_2 and are computed from (5.25), (5.26) and (5.27) and then obtained as [7]:

$$\hat{x}_1 = (h_1^2 + h_2^2)x_1 + \overline{w}_1 \tag{5.28a}$$
$$\hat{x}_2 = (h_1^2 + h_2^2)x_2 + \overline{w}_2 \tag{5.28b}$$

As the noise samples $\overline{w}_1$ and $\overline{w}_2$, channel gains h_1 and h_2 are independent and uncorrelated, the two reconstructed signals are orthogonal and uncorrelated. Hence, the detection problem for u_1 and u_2 decomposes into two separate, orthogonal, scalar problems. Therefore, like in the case of SISO ML detection can be used for decision. Thus, the optical Alamouti provides the diversity gain of two for the detection of each symbol. This is possible because the two symbols are now transmitted over two symbol times instead of one symbol, unlike in the case of repetition coding. But each symbol has only half the power, assuming that the total transmitted power is the same in both cases. Figure 5.6 gives the BER performance of the optical Alamouti with OOK modulation for the Log-Normal distributed fading channel.

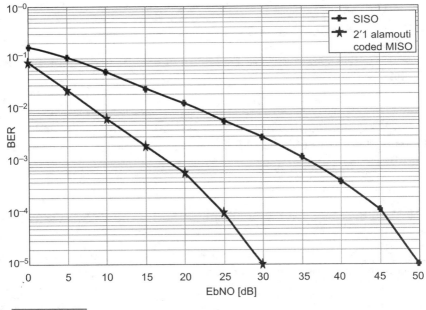

Figure 5.6   **BER Performance of 2x1 Alamouti for OOK with Log-Normal Channel.**

5.6 MIMO CHANNEL

Finally, we consider the optical MIMO channels. MIMO channels are frequently used in RF wireless system for diversity, array, as well as multiplexing gains [8-10]. In a MxN MIMO system, there are M transmit and N receive antennas (Fig. 5.5(c)) with h_{mn} as a randomly distributed channel gain from transmit antenna m to receive antenna n. The fading gains h_{mn}'s are assumed to be independent as the transmit antennas as well as the receive antennas are spaced sufficiently far apart. There are several aspects of optical MIMO, which are different from their RF counterparts and, therefore, same approaches cannot be applied directly. We know that channel optical input symbols are not complex numbers but are non-negative real intensity signals and so are all the h_{mn} entries of the channel matrix. Also, the received signal at times has to be considered as Poisson instead of Gaussian in the power limited highly sensitive systems as discussed in Section 5.2.

In the following sections, we first consider the repetitive diversity MIMO systems and then take up spatial multiplexed MIMO systems.

5.6.1 Repetitive MIMO Systems

For the $M \times N$ MIMO system, as we are assuming the separation among the M sources as well as the N detectors to be more than the coherence distance, the fading experienced between source-detector pairs is statistically independent. Therefore, if same symbol (repetitive) is sent from all the transmitters, each symbol experiences different fading, therefore, the diversity benefits can be accrued from the MIMO channel. On each small detector one can model the received field as spatially constant though randomly distributed over detector aperture and the path gains from any one source to various detectors are independent. The channel fading process is assumed to be flat across the optical frequency band and slow relative to the symbol duration.

Considering once again the OOK transmission, we first focus on the Poisson regime [11], which, as mentioned earlier, is applicable in the low SNR region. Each of the N receivers consists of a photodetector and integrator as shown in Fig. 5.3. The optical signal is received at the detector and the output is a flow of photoelectrons that obey Poisson statistics over the bit interval.. If the effective count on the nth receive antenna in the absence of fading for the *On* bit be $n_{on,n}$ and for *Off* bit the count be $n_{off,n}$, then from (5. 15) these are given by

$$n_{on,n} = \frac{\eta P_s T_b}{hfM} + \frac{\eta P_b T_b}{hf} \qquad (5.29a)$$

and

$$n_{off,n} = \frac{\eta P_b T_b}{hf} \qquad (5.29b)$$

for n = 1, 2, ...N. P_b is the optical power received due to background radiation and P_s is the power provided at one detector by the entire transmitter array. The variance of the observable, i.e., the count at the nth detector for the qth bit, $Z_n(q)$ is a Poisson count variable and it has the average value expressed as,

$$n_{on,n} s[q] G_n + n_{off,n} \qquad (5.30)$$

where, $s[q] \in \{0,1\}$ is the OOK data symbol and $G_n = \sum_{m=1}^{M} h_{m,n}$ is the fading intensity at the nth detector, which is defined as the sum of channel fading for the links from m^{th} transmitter to n^{th} detector and considered to be constant over the bit interval q. Representing the counts on the detector by a matrix $Z = \{Z_n(q), n = 1,2,...N\}$, the probability mass function of the Poisson variate, $Z_n(q)$ can be written as:

$$\Pr\{Z_n[q], G_n\} = \frac{(n_{\text{on},n}s[q]G_n + n_{\text{off},n})^{Z_n[q]} e^{-(n_{\text{on},n}s[q]G_n + n_{\text{off},n})}}{Z_n[q]!} \tag{5.31}$$

With the receiver having the complete statistical information of the channel denoted by the channel state information (CSI), we can express the optimal ML detection as:

$$\frac{\prod_{n=1}^{N} \Pr\left\{Z_n\{q\}, G_n; s[q] = 1\right\}}{\prod_{n=1}^{N} \Pr\left\{Z_n\{q\}, G_n; s[q] = 0\right\}} \overset{\hat{s}[q]=1}{\underset{\hat{s}[q]=0}{\gtrless}} 1 \tag{5.32}$$

Here, the sum of all faded signals coming from different channels are intensity signals and, therefore, is always constructive in the case of FSO.

To simplify the receiver structure, instead of comparing signal of each detector, one can use equal gain combining (EGC) of the signals at the N detectors before the decision of the *On* or *Off* bit is taken. The EGC statistics for the qth bit interval $R[q]$, is defined as $\sum_{n=1}^{N} Z_n[q]$. This will also be Poisson distributed and the detection with EGC will be then expressed as:

$$\frac{\Pr\left\{R[q], H; s[q] = 1\right\}}{\Pr\left\{R[q], H; s[q] = 0, H\right\}} \overset{\hat{s}[q]=1}{\underset{\hat{s}[q]=0}{\gtrless}} 1 \tag{5.33}$$

where, H is the overall channel matrix for the MIMO system defined as $\sum_{n=1}^{N} G_n$. The receiver structure is shown in Fig. 5.7. This simplified receiver gives a reasonably good accuracy.

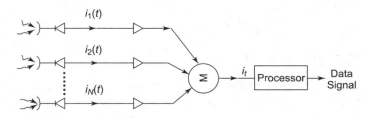

Figure 5.7   **The Receiver Structure of Repetitive MIMO.**

Next, we consider the case of detection in the *Gaussian regime*. This is valid for high energy systems, and when the shot-noise in the data signal is negligible relative to the background radiation. The power levels of the background and signal are such that the Gaussian approximation is justified. In this case, we can state that the observable $Z_n[q]$, which is the output voltage across the resistance, is a Gaussian random variable with the probability mass function for the *On* or *Off* bit given respectively as:

$$\Pr\{Z_n[q], On\} = \frac{1}{\sqrt{2\pi\sigma_{on,n}^2}} \exp\left(-\frac{(Z_n[q]-\mu_{on.n})^2}{2\sigma_{on,n}^2}\right) \tag{5.34a}$$

and

$$\Pr\{Z_n[q], Off\} = \frac{1}{\sqrt{2\pi\sigma_{off,n}^2}} \exp\left(-\frac{(Z_n[q]-\mu_{off.n})^2}{2\sigma_{off,n}^2}\right) \tag{5.34b}$$

The mean and variance parameters for *On* and *Off* signal are given as:

$$\mu_{on.n} = e\left(\frac{\eta P_s T_0}{hf}G_n + \frac{\eta P_b T_0}{hf}\right)R_L \tag{5.35a}$$

$$\mu_{off} = \frac{e\eta P_b T_0}{hf}R_L \tag{5.35b}$$

$$\sigma_{on.n}^2 = e(\mu_{on,n})R_L + 2kT_b T_0 R_L \tag{5.36}$$

$$\sigma_{off}^2 = e(\mu_{off})R_L + 2kT_b T_0 R_L \tag{5.37}$$

where, R_L is the load resistance across the detector and T_0 the integration period. Similar to the development for the Poisson regime, the log likelihood function for ML detection with Gaussian distribution is the expressed as:

$$\Lambda(q) = \sum_{n=1}^{N}\left\{\left(\frac{Z_n[q]-\mu_{on,n}}{\sigma_{on,n}}\right)^2 - \left(\frac{Z_n[q]-\mu_{off,n}}{\sigma_{off,n}}\right)^2\right\} \tag{5.38}$$

Figure 5.8 gives the performance of repetitive MIMO system with 2-PPM modulation scheme [11] with Poisson channel.

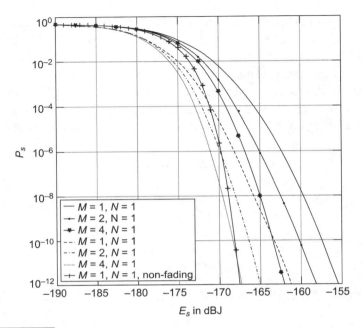

E_s in dBJ

Figure 5.8 BER vs. Received Optical Energy Es Per Symbol in the Absence of Fading for Binary PPM for Repetitive MIMO and Scientillation Index of 0.4 [11].

5.6.2   MIMO Channel With Spatial Multiplexing

In the case of MxN MIMO channel with spatial multiplexing, each transmitter transmits different symbols simultaneously towards the array of N detectors. In high turbulent atmosphere with channel gains from different transmit to receive antenna being uncorrelated, multiple data independent channels open up which contribute to multiplexing gain in the link. All the transmitted symbols are received by each of the N receive antennas. In each use of the MIMO channel, a vector $\bar{x} = (x_1, x_2, ...x_M)^T$ of real numbers is sent and a vector $\bar{y} = (y_1, y_2, ...y_N)^T$ of real numbers is received. The input-output relationship of the system can be expressed as:

$$\bar{y} = H\bar{x} + \bar{w} \tag{5.39}$$

where, H is a $N \times M$ matrix representing the scattering effects of the channel and $\bar{w} = (w_1, w_2, ...w_N)^T$ is the Gaussian noise random vector with i.i.d. elements. H is a random matrix with independent real elements $\{h_{mn}\}$, and H and w are assumed to be independent of each other and also of the data vector x. It is assumed that the receiver has perfect knowledge of the channel realization H, while the transmitter has no such channel state information.

In the MxN MIMO, there are now maximum of MN independently faded signal paths between the transmitter and the receiver. Therefore, if we once again use the same repetition scheme, i.e., transmitting the same symbol over the M antennas, which are received independently through the atmospheric channel by each of the N receive antennas, then one can achieve a maximum diversity gain of MN,. The MN received symbols can be combined to obtain a MN-fold diversity gain and an effective channel gain of $\sum_{m=1}^{M} \sum_{n=1}^{N} \left| h_{mn} \right|^2$. However, just as in the case of the MISO channel, the degrees of freedom of the channel are utilized poorly in the repetition scheme; it only transmits one data symbol per M symbol times. The *degrees of freedom (DoF)* can be understood to be the number of independent signals received by the antenna after propagating over the channel. For a SIMO system though the received signal lies in an N-dimensional vector space, but as all the antennas receive the same signal through the turbulent atmosphere, the degrees of freedom is still *one* per symbol time. For the case of MISO with M transmit and one receive antenna, it is equal to one for every symbol time. The repetition scheme utilizes only $1/M$ of the degree of freedom per symbol time. But in the case of (2x1) Alamouti scheme, it is *one* instead of 1/2. On the other hand, if we consider a 2 x 2 MIMO channel transmitting two independent uncoded symbols over the two different antennas as well as over the different symbol times a complete utilization of channel *DoF* of *two* can be achieved. This is the spatial multiplexing scheme gain. If the channel gains are linearly independent, the signal space dimension is 2; the signal from say, transmit antenna 1 arrives along h_{11} or h_{12}, and from transmit antenna 2 arrives along h_{21} or h_{22} at the two receive antennas, the receiver can distinguish between the two signals coming from two transmit antennas. Therefore, compared to a 2 x 1 channel, there is an additional degree of freedom coming from space. The channel capacity, therefore, increases in spatial multiplexing. A block diagram for the spatially multiplexed MIMO is shown in Fig. 5.9.

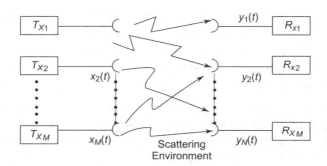

Figure 5.9   Block Diagram of Spatially Multiplexed MIMO.

In the spatially multiplexed MIMO system with the increase of channel capacity the detection complexity also increases. The complexity in detection is due to the fact that the signals launched from the different transmit antennas interfere with each other as different data streams are transmitted occupying the same resources in time and wavelength. The signal detection problem in spatially multiplexed MIMO channel is the problem of estimating the input vector $\bar{x}$ given the received vector $\bar{y}$ assuming that the receiver has perfect knowledge of H.

To estimate the transmitted signal from the received signal, we first exclude the coding across the time dimension. The decision is then made on a symbol by symbol basis without taking into account any statistical dependencies that may be present in the sequence of vectors $\bar{x}$. Each element of $\bar{x}$ is assumed to belong to a common modulation alphabet X, with $x_m \in X$, $m = 1,\ldots\ldots M$. Therefore, the random vector $\bar{x}$ can be expressed as:

$$E\{xx^T\} = \frac{E_T}{M} I_M \tag{5.40}$$

where, I_M is the identity matrix of size M, E (.) is the expectation operator, and x^T denotes the transpose of x. Assumption (5.40) implies that the elements of $\bar{x}$ are uncorrelated and each has energy $\dfrac{E_T}{M}$, yielding a total average transmitted energy of E_T per symbol combined over all antennas. The average received energy per symbol E_s at each receiver antenna is also equal to E_T, which can be seen by computing the energy as:

$$Es = E\left\{ \left| \sum_m h_{nm} x_m \right|^2 \right\}$$

$$= \sum_m \sum_k E(h_{nm} h_{nk}^T) E(x_m x_k^T)$$

$$= \sum_m E(|x_m|^2) = E_T \tag{5.41}$$

In the above equation, h_{nm} has been assumed to be log-normal with mean zero and variance of 1. Using the above equation, the average received energy per bit at each receiver antenna may be computed as:

$$E_b = \frac{E_s}{\log_2 |X|} \tag{5.42}$$

and receiver SNR is then defined as:

$$\text{SNR} = \frac{E_b}{N_o} = \frac{E_T / \log_2 |X|}{N_o} \tag{5.43}$$

To minimize the probability of error, the MAP decision rule [1] can be applied to obtain the transmitted signal as:

$$\hat{x} = \arg \max_{x \in X^M} \left[P_r(x|y) \right] \tag{5.44}$$

On the other hand, the ML rule for all equally likely source symbols to be transmitted a-priori, can be expressed as:

$$\hat{x} = \arg \min_{x \in X^M} \left| y - Hx \sqrt{\frac{E_s}{M}} \right|^2 \tag{5.45}$$

Although MAP/ML rule offers optimal error performance, it suffers from complexity issues in the case of spatially multiplexing. The complexity of the receiver increases exponential as the number of antennas at the transmitter increases. This is because the receiver has to consider $|X|^M$ possible symbols for an M transmitter antenna system. For example, for 16PPM with 4x4 MIMO the MAP or ML detector needs to search over 16^4 symbols. The difficulty of detection is caused as each antenna observes a superposition of the transmitted signals, which gets more complicated due to multiplication by H.

As the optimum detection in MIMO systems is quite complex with ML/MAP, sub-optimal detection models are used. These give good error performance but still remain practically tractable as far as implementation complexity is concerned. One such receiver is the V-BLAST (Vertical Bell-Labs Layered Space-Time) receiver which utilizes a layered architecture and applies successive cancellation by splitting the channel vertically [12-14]. This is a class of linear receivers. *Linear receivers* separate the transmitted data streams and then independently decode each of the streams. The symbol estimate of $\hat{x}$ is given by a transformation of the received vector $\bar{y}$ of the form:

$$\hat{x} = Q(Wy) \tag{5.46}$$

where, Q is a quantizer, that maps its argument to the nearest signal point in X^M using Euclidian distance, and W is the weighting matrix that may depend on H. The linear receivers are of the following types:

➤ *Zero-Forcing (ZF) Detection:* These receivers use the low-complexity zero-forcing linear detection algorithm and outputs the estimate of the transmitted signal as:

$$\hat{x} = Q(\hat{x}_{ZF}) \tag{5.47}$$

where,

$$\hat{x}_{ZF} = H^+ y \tag{5.48}$$

and H^+ denotes the Moore-Penrose pseudoinverse [15] of H, which is a generalized inverse that exists even when H is rank-deficient. The estimated output is therefore:

$$\hat{x} = x + \sqrt{\frac{M}{E_s}} H^+ y \tag{5.49}$$

From equation (5.49) we see that the matrix channel is decoupled into M parallel scalar channels with additive but noise is correlated. Each channel is decoded independently ignoring noise correlation across the processed streams. The receiver complexity is reduced considerably by thus converting the joint-decoding problem into M single stream-decoding problem. The ZF receiver thus eliminates the co-channel interference entirely since $H^+H = I$, but in the process enhances the noise power. The diversity order achieved by the individual data streams is N-M+1 [8].

➤ *Linear Least Squares Estimation (LLSE) Detection:* In the case of LLSE detection algorithm the output estimation is:

$$\hat{x} = Q(\hat{x}_{LLSE}) \tag{5.50}$$

where, $(\hat{x}_{LLSE})$ is a linear estimator given by:

$$(\hat{x}_{LLSE}) = Wy \tag{5.51}$$

where, W is chosen to minimize the following difference:

$$E\{\|Wy - x\|^2\} \tag{5.52}$$

The LLSE estimator matrix W can be given by [8]:

$$W = \frac{E_T}{M} H^T \left(\frac{E_T}{M} HH^T + N_o I_N \right)^{-1} \tag{5.53}$$

In the case of LLSE estimator the co-channel interference is not eliminated entirely, but on the other hand, it has the desirable property of not enhancing noise as much as in the case of the ZF estimator. In the low-SNR regime it approaches the matched-filter receiver and outperforms the ZF detector. At high SNR ($E_b/N_0 \gg 1$) the LLSE receiver approaches the ZF receiver and the diversity order achieved by the individual data streams is N-M+1. [8].

➤ *V-BLAST Detection:* The V-BLAST detection algorithm [12] is a recursive procedure that extracts the components of the transmitted vector x according to a certain ordering $(k_1, k_2 \ldots k_M)$ of the indices of the elements of $\bar{x}$. The permutation of $(k_1, k_2 \ldots k_M)$ depends on H as the channel state information is assumed to be known at the receiver. In V-BLAST the strongest signal is first decoded rather than jointly decoding all the transmit signals. Next, this signal is subtracted from the received signal and further proceeded to decode the strongest signal of the remaining transmit signals thereon. The optimum detection order is, therefore, a *nulling and cancellation* strategy proceeding from the strongest to the weakest signal. The V-BLAST algorithm is as follows:

1. *Nulling:* An estimate of the strongest transmit signal is obtained by nulling out all the weaker transmit signals using either ZF or LLSE or MAP criterion.

2. *Slicing:* The estimated signal is detected to obtain the data bits.

3. *Cancellation:* These data bits are re-modulated and the channel is applied to estimate its vector signal contribution at the receiver. The resulting vector is then subtracted from the received signal vector and the algorithm returns to the nulling step until all transmit signals are detected.

The V-BLAST algorithm can have different variants, i.e., V-BLAST/ ZF; derived from ZF rule in which the successive-cancellation scheme is derived from the ZF scheme discussed above. Similar to V-BLAST/ZF one may also have V-BLAST/LSSE or V-BLAST/ZF/MAP. The performance of the MIMO Gaussian channel with VBLAST/ZF is given in Fig. 5.10 [16].

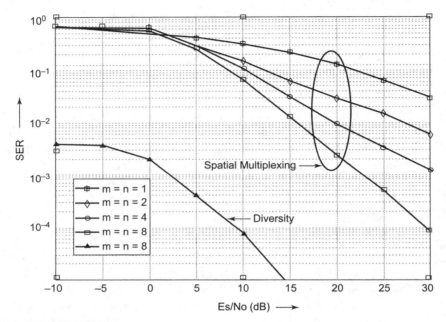

Figure 5.10   The BER Performance with VBLAST/ZF with PPM.

TIME DIVERSITY

In order to increase the BER performance of the FSO link, instead of using multiple antennas at the transmitter or receiver end which increases the complexity and/ or cost, one can work with a single beam at the transmitter and a single lens at the receiver by using time diversity. In the time diversity scheme, identical messages are transmitted in different time slots separated by time periods of the order of the coherence time [17]. Statistically, laser propagation along an atmospheric path is uncorrelated with an earlier-time path for a time interval greater than the atmospheric turbulence correlation time. Typically, the channel coherence time is of the order of 100's of symbols, and therefore the channel is highly correlated across consecutive symbols as the channel characteristics do not change over the coherence time.

The block diagram for a time diversity system is given in Fig. 5.11. The transmitter sends an intensity signal L times, separated by fixed time periods that are much larger than typical deep fade duration. The receiver receives independently faded copies of the signal, combines them using a combining scheme, and determines whether a 0 or 1 is sent.

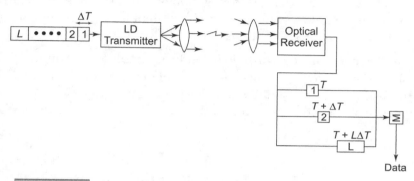

Figure 5.11   FSO System Block Diagram with Time Diversity.

For simplicity, let us consider a flat fading channel. The transmitted signal of length L symbols is expressed as $\bar{x} = [x_1,.....x_L]^t$ and the received signal is given by

$$y_l = h_l x_l + w_l; \text{ with } l = 1,2,...L \tag{5.54}$$

We can assume that the h_l's are independent by assuming ideal interleaving, i.e., the consecutive symbols x_l are transmitted sufficiently far apart in time. The parameter L is commonly called the *number of diversity branches*. The additive noises w_1, ..., w_L are i.i.d. AWG random variables. The receiver thus receives L independently faded copies of the same signal, applies appropriate delay to each copy, combines them, and demodulates the messages. When each slot is longer than the correlation time of the atmosphere, the bit errors reduce. Communication performance is improved because the joint probability of error is less than the probability of error from individual channels [18]. In the case of burst-mode data transmission when the data frame length is large compared to the channel coherence interval, one can exploit the time diversity by employing channel coding and interleaving.

For the detection of time diversity signals, Taylor frozen turbulence hypothesis can be used. Accordingly, the spatial statistics of the atmosphere can be converted to temporal statistics by the knowledge of the average wind speed transverse to the direction of propagation expressed as [6]:

$$n_1(r,t) = n_1 (r - Vt, 0) \tag{5.55}$$

where, V is the wind velocity, which is equal to the average wind velocity transverse to the direction of propagation and in the plane of the array when the fluctuation in the velocity are negligible, n_1 is the fluctuating component of the refractive index.

The detection in the turbulence-induced fading is computed by threshold detection of the joint distribution of the received signal intensity for the sequence of transmitted bits. The joint distribution of intensity for a sequence of transmitted *On* bits at the receiver can be expressed as:

$$f_I(I_1...I_L) = \frac{1}{2^L (2\pi)^{L/2} \left| B_X^{\tau_0} \right|^{1/2} \prod_{l=1}^{L} I_l} \exp\left\{ -\frac{1}{8} \left[\ln\left(\frac{I_1}{I_0}\right) ... \ln\frac{I_L}{I_0} \right] . (B_X^{\tau_0})^{-1} . \begin{bmatrix} \ln\left(\frac{I_1}{I_0}\right) \\ \ln\left(\frac{I_L}{I_0}\right) \end{bmatrix} \right\} \tag{5.56}$$

where, $B_X^{\tau_0}$ is the auto-covariance matrix of log-amplitude fluctuation X, at different time [6] for a string of L bits with τ_0 as the bit interval greater than the correlation time. This is expressed as:

$$B_X^{\tau_0} = \sigma_X^2 \begin{bmatrix} 1 & b_X(V\tau_0) & \cdots & b_X[V(L-1)\tau_0] \\ b_X(V\tau_0) & & \cdots & b_X[V(L-2)\tau_0] \\ \cdots & \cdots & \cdots & \cdots \\ b_X[V(L-1)\tau_0] & b_X[V(L-2)\tau_0] & \cdots & 1 \end{bmatrix} \tag{5.57}$$

where, σ_X^2 is the variance of X. The log-amplitude turbulence induced fluctuation X is related to the estimated received light intensity signal I as:

$$I = I_0 \exp(2X - 2E[X]) \tag{5.58}$$

In (5.57) the correlation time is defined as the $1/e^2$ point of the normalized covariance function $b_X(\tau_0)$ of X for bit interval of τ_0 which is given by:

$$b_X(\tau_0) = \frac{B_X(\tau_0, D)}{B_X(0, D)} \tag{5.59}$$

where, B_X is the log-amplitude temporal covariance function defined as [19]:

$$B_X(\tau_0, D) = 8\pi^2 k^2 L_T \int\limits_0^1 \int\limits_0^\infty \kappa \phi_n(\kappa) J_0(\kappa V \tau_0) x \exp\left(-\frac{D^2 \kappa^2}{16}\right)\left(1 - \cos\frac{L_T \kappa^2 \xi}{k}\right) d\kappa d\xi$$

(5.60)

where, D is the receiver diameter, ϕ_n is the spatial power spectrum of refractive index, J_0 is the Bessel function of the first kind, L_T is the propagation distance, κ is the spatial wave number and ξ is the transformation variable.

The probability of fade at the receiver, thus can be obtained from the joint distribution of intensity for a sequence of transmitted bits as:

$$P(I < I_{th}) = \int\limits_0^{I_{th}} ... \int\limits_0^{I_{th}} f_I(I_1...I_L) dI_1 dI_2 ... dI_L$$

(5.61)

Though one can potentially benefit from time diversity but the delay period to obtain full time diversity gains are of the order of few milliseconds and hence will increase the communication latency and reduce the bit rate. Benefits from time diversity signal can also be accrued by coding. The repetition code is the simplest possible code. Although it achieves a diversity gain, it does not exploit the degrees of freedom available in the channel effectively because it simply repeats the same symbol over the L symbol times. By using more sophisticated codes, to be discussed in Chapter 7, a coding gain can also be obtained beyond the diversity gain.

Summary

We first looked at the detection problem in a turbulent random channel. The error probability was found to be linearly varying with inverse of SNR at high SNR. In contrast, the error probability decreases *exponentially* with the SNR in the AWGN channel discussed in Chapter 3. The typical error event for the fading channel is due to the channel being in a deep fade rather than the Gaussian noise being large. Therefore, diversity is required to improve the performance. Next, diversity was described, which is an effective approach to improve performance drastically by providing redundancy across independently faded branches.

Two modes of diversity were considered:

- Space diversity with the use of multiple transmit and/or receive antennas. In MISO channel, better performance is obtained with space-time Optical Alamouti code. But both, MISO and SIMO systems fail to provide the complete use of all the DoF though they do provide array and diversity gain. With MIMO, spatial multiplexing increases the channel capacity linearly with the number of antennas. The detection of the received signals becomes increasingly complex in this case with optimum ML/ MAP detection. Sub-optimal detection scheme V-BLAST is resorted with minimal complexity and fairly reasonable error performance.

- Time diversity by transmitting repetitive symbols over different coherence time periods or with burst-mode data transmission when the data frame length is large compared to the channel coherence interval by employing channel coding and interleaving.

Use of wavelength diversity has not been applied to FSO systems practically, though some studies have been carried out to use multiple frequencies to mitigate the effect of fog [20]. Purvinskis *et al.*[21] have also used multiple wavelengths to alleviate the effect of atmospheric turbulence.

References

1. Proakis J.G., *Digital Communications*, McGraw-Hill, 5th edition, 2008.
2. Theodore Rappoport; *Wireless Communications: Principles and Practice*, Prentice Hall, 2nd edition, 2001.
3. John M. Senior, *Optical Fiber Communication*, Prentice Hall Publication (UK), 2nd edition, 1993.
4. Max Born and Emil Wolf, *Principles of Optics*, 6th edition, Pergamon Press, Canada, 1980.
5. Papoulis A., *Probability, Random Variables & Stochastic Processes*, McGraw-Hill, 3rd edition, 1991.
6. Zhu X. and Kahn J.M., 'Free-Space Optical Communication through Atmospheric Turbulence Channels', *IEEE Trans. Commun.* **50** (8), August 2002, 1293–1300.
7. Simon M.K. and Vilnrotter V.A., ' Alamouti-type space-time coding for free-space optical communications with direct detection,' *IEEE Trans. Wireless Comm.*, Vol 4, Jan 2005, pp 35-39.
8. Bolcskei H. and Paulraj A.J., 'Multiple-input multiple-output (MIMO) wireless systems', pp. 90.1–90.14. Edited by J Gibson, *The Communication Handbook*, CRC Press, 2nd ed., 2002.

9. Paulraj A.J., Gore D., Nabar R.U. and Bolcskei H., 'An overview of MIMO communications—A key to gigabit wireless', *Proc. IEEE*, Vol. 92, No. 2, pp 198–219, Feb. 2004.

10. George Ed., Tsoulas, *MIMO System Technology for Wireless Communications*, CRC Tailor & Francis, 2006.

11. Wilson S., Brandt-Pearce M., Leveque J. and Cao Q., 'Free-space optical MIMO transmission with Q-ary PPM', *IEEE Trans Comm*, Vol. 53, no. 8, Aug. 2005

12. Wolniansky P.W., Foschini G.J., Golden G.D., and Valenzuela R.A., 'V-blast: An architecture for realizing very high date rates over the rich scattering wireless channel', *Proc. URSI ISSSE*, pp. 295–300, 1998.

13. Foschini G. J., 'Layered space-time architecture for wireless communication in a fading environment when using multiple antennas', Bell Labs Tech. J., Vol. 1, pp. 41–59, Autumn 1996.

14. Foschini G.J. and Gans M.J., 'On limits of wireless communications in a fading environment when using multiple antennas,' Wireless Pers. Comm., Vol. 6, no. 3, pp. 311–335, 1998.

15. Golub G.H. and Loan C.F.V., *Matrix computations*, John Hopkins University Press, Baltimore, 1983.

16. Sandhu H. and Chadha D., 'Power and spectral efficient free space optical link based on MIMO system', CNSR 2008, Comm, Networks and Service Resources, May 5th–8th, 2008, Halifax, Canada.

17. Fang Xu, Ali Khalighi, Patrice Causs'e, and Salah Bourennane 'Channel coding and time-diversity for optical wireless links' Optical Express 17, 872–887, 2009.

18. Ibrahim M.M. and Ibrahim A.M., 'Performance analysis of optical receivers with space diversity reception,' *Proc. IEE—Comm.*, Vol. 143, no. 6, pp. 369–372, Dec 1996.

19. Andrews L.C. and Phillips R.L., *Laser Beam Propagation through Random Media*, 2nd ed., SPIE Press, Bellingham, Washington, 2005.

20. Eric Wainright, Hazem Refai H., and James Sluss J., *Wavelength Diversity in Free-Space Optics to Alleviate Fog Effects*, Free-Space Laser Comm Technologies XVII, edited by G. Stephen Mecherle, *Proc. of SPIE*, Vol. 5712, 2005, pp 110–118.

21. Purvinskis et al, 'Multiple wavelength free-space laser comm,' *Proc. of SPIE*, Vol. 4975, 2003.

Chapter 6

Channel Capacity

Due to turbulence in the atmosphere, when the signal is in deep fade, there is a considerable deterioration in the performance of the signal in the free space optical (FSO) channel. As discussed in Chapter 5, unlike in the case of AWGN channel, the situation in the FSO system cannot be improved by increasing the signal power only; instead diversity is to be used to improve the BER performance of the signal in the low SNR region. Also, the error performance can be improved in the lower SNR region by using advanced error correcting coding schemes, such as, the Turbo or Low Density Parity Check codes so as to approach the optimum performance. In the present chapter, we discuss about this optimal performance achievable in a FSO fading channel. One of the basic measures of optimal performance is the *capacity* of a channel defined as the maximum rate of communication for which an arbitrarily small error probability can be achieved. Before Shannon [1], it was thought that the only way to make the error probability as small as desired over a noisy non-ideal channel is to reduce the data rate by repeatedly sending the same bits, which is basically repetition coding. It was suggested by Shannon that, in fact, by coding the information appropriately, one can communicate at a strictly positive rate and also at the same time with as small an error probability as desired. There is a maximal rate of the channel for which this can be done. Any rate above the channel capacity will drive the system to a possible higher error probability. We discuss several different definitions of channel capacity available in literature through which the capacity of optical fading channel is explained. These measures also explain the resources available in fading channels to increase its capacity, such as the power, bandwidth and degrees of freedom.

As in Chapter 5, starting with the AWGN channel as the building block, we first calculate the capacity of the AWGN channel and determine the resources available to increase the capacity in section 6.1. With the background of this non-fading AWGN channel, the capacity of optical wireless fading channels are discussed for Gaussian detection in the case of SISO links in Sections 6.2 and 6.3. Next, channel capacity of multiple transmit and receive antenna for FSO has been considered both as spatially multiplexed and repetitive MIMO systems in Section 6.4. Lastly, the channel capacity of low power ideal Poisson channel is discussed both for SISO and MIMO system in Section 6.5.

6.1　CHANNEL CAPACITY OF AWGN CHANNEL

The AWGN optical channel without fading at discrete time instant can be modeled by the base band electrical signal level as:

$$y[k] = x[k] + w[k] \tag{6.1}$$

with $x[k]$ and $y[k]$ as the real input and output signals at time instant k, respectively and $w[k]$ as the additive white gaussian noise with a variance of σ^2 independent over time. The input optical signal has the constraints of electrical signal x to be positive and an average amplitude limited by an optical power due to eye/skin effects. .

Consider the case of OOK intensity modulation. As derived in Chapter 4, the symbols $x[k]$ take values of 0 and $2P$ and the error probability achievable with optimum detection is $Q\left(\dfrac{\sqrt{T_b}}{\sigma}P\right)$. The issue is how can we reduce this error? Simple and somewhat obvious way is to send the same bit of information repeatedly K times. The code-words for '1' and '0' bits of this repetition coding will be, $x_A = 2P[1,\dots1]^t$ and $x_B = [0,\dots0]^t$, respectively with an average power constraint of P.

Now the received vector, when x_A is transmitted over the AWGN channel is:

$$y = x_A + w \tag{6.2}$$

where, $w = (w[1],\ldots\ldots,w[K])^t$.

The error will occur for this case whenever y is closer to x_B than to x_A, with the error probability given by

$$Q\left(\frac{\|x_A - x_B\|}{2\sigma}\right) = Q\left(\frac{KP\sqrt{T_b}}{\sigma}\right) \tag{6.3}$$

We observe from the above expression that the probability of error decays exponentially with the increase of block length K. Hence the communication can now be done with arbitrary reliability by choosing a large value of K, but the price paid is reduced data rate of $1/K$ bits per symbol time. Increasing K will make the data rate go to very small value. So, what is the maximum data rate that can be transmitted without error over an AWGN channel with a given power? This can be explained by the classical *sphere-packing* theorem. By the law of large numbers, the n-dimensional received vector $y = (x + w)$ will lie with high probability within a y-sphere of radius $\sqrt{n(P_s + P_N)}$ [2], where P_s and P_N are the received electrical signal power and the noise power, respectively. By the law of large numbers, the noise power is given as:

$$\frac{1}{n}\sum_{k=1}^{n} w_i^2[k] \to P_N \text{ as } n \to \infty \tag{6.4}$$

So as the number of transmitted signals increase, the received vector y lies with high probability near the surface of a noise sphere of radius $\sqrt{nP_N}$ around the transmitted codeword. Therefore, as long as the noise spheres around the symbol do not overlap so that the received signal does not make a wrong decision, reliable communication can take place. This gets explained in Fig. 6.1. Hence, in general, the maximum number of symbols that can be packed with non-overlapping n-dimensional noise spheres is given as the ratio of the volume of the n-dimensional hypersphere to the volume of a noise sphere:

$$\frac{\left[\sqrt{n(P_s + P_N)}\right]^n}{\left[\sqrt{nP_N}\right]^n} \tag{6.5}$$

This implies that the maximum number of bits per symbol that can be reliably communicated is

$$\frac{1}{n}\log\left(\frac{\{n(P_s + P_N)\}^{n/2}}{(nP_N)^{n/2}}\right) = \frac{1}{2}\log\left(1 + \frac{P_s}{P_N}\right) \tag{6.6}$$

This is the capacity of the AWGN channel. In other words, this gives the maximum symbol rate for reliable communication in an optical AWGN channel with the required constraints on x. In practical systems, the bit rate is to be kept lower than the capacity of the channel.

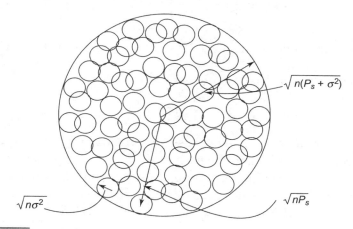

Figure 6.1 The Received Signal Sphere with the Noise Signal Spheres.

We know that repetition coding discussed above is not an optimum code. It is very inefficient as it uses only one dimension out of the large number of dimensions of the signal space. There are a few optimum codes for FSO systems discussed in the next chapter which can enhance the capacity of the FSO fading channel considerably.

Next, we look into the resources, which are available to increase the channel capacity. There are three resources; *power, bandwidth,* and *degree of freedom,* which can increase the capacity of the channel. In the following paragraphs, first, we see the role played by these resources in the AWGN channels before we discuss the channel capacity for the optical fading channels.

Consider a continuous-time AWGN channel with bandwidth W Hz, average power constraint P Watts, and additive white Gaussian noise with power spectral density of $N_0/2$. By sampling the received signal at rate $1/W$, the channel can be represented by the discrete-time baseband channel as given in (6.1). Hence, from (6.6) the capacity of the channel for a symbol transmission is:

$$C = \frac{1}{2}\log\left(1 + \frac{P_s}{N_0 W}\right) \text{ bits} \tag{6.7}$$

Since there are W real samples per second, the capacity of the continuous-time AWGN channel is:

$$C_{awgn} = \frac{W}{2}\log\left(1 + \frac{P_s}{N_0 W}\right) \text{ bits/s} \tag{6.8}$$

As $\dfrac{P}{N_0 W}$ is the SNR per degree of freedom, the spectral efficiency of the AWGN channel can also be expressed as:

$$= \frac{1}{2}\log(1 + SNR) \text{ bits/s/Hz} \tag{6.9}$$

From equation (6.8), we note the basic resources available in the channel are the *received powe P_s*, and the *bandwidth, W*. It does not depend on the modulation scheme used. Therefore, without going into the details of specific modulation schemes used one can use the capacity formula (6.8) to evaluate the performance of a communication system.

Let us first see how the capacity depends on the received power. From equation (6.9), we observe that the capacity is logarithmically proportional to SNR. When the SNR is low, the capacity increases linearly with the received power, but when the SNR is high, the capacity increases logarithmically with received power. Therefore, increasing the power P_s will increase the capacity, but will suffer from a Law of Diminishing Marginal Returns. The higher the SNR or power for fixed noise, the smaller the effect on capacity.

The capacity depends on the bandwidth in two ways [3]. The capacity has a distinctive variation in the two regions of low and high bandwidth. In the low bandwidth region, when the bandwidth value is small, from (6.8) the capacity can be seen to be linearly dependent on W for a fixed SNR. In this region when the bandwidth is small, the SNR per degree of freedom is high, and therefore, the capacity is insensitive to small changes in SNR. But as the bandwidth increases in this region, the degrees of freedom available for communication also increase; in other words, more samples/sec can be transmitted, and therefore a fast increase in the capacity occurs. The increase in the degrees of freedom more than compensates for the decrease in SNR, which occurs with the increased bandwidth because with increasing bandwidth though more information of signals will be transmitted per sec but along with it, the noise will increase. The system is, therefore, said to be in the *bandwidth-limited regime*. On the other hand in high bandwidth region with increasing bandwidth for a given received power P_s, the SNR *per dimension* decreases as the energy is spread more sparsely across the degrees of freedom. When the bandwidth is large such that the SNR per degree of freedom is small the spectral efficiency reduces to,

$$\log\left(1+\frac{P_s}{N_0W}\right) \approx \left(\frac{P_s}{N_0W}\right)\log_2 e \text{ bit/s/Hz} \tag{6.10}$$

In this regime, the capacity is proportional to the total received power across the entire band. The capacity is now linear in the received power. This is the *power-limited regime*. As W increases, the capacity increases monotonically and reaches the asymptotic limit of:

$$C_\infty = \frac{P_s}{N_0}\log_2 e \text{ bits/s} \tag{6.11}$$

This is the infinite bandwidth limit; in other words, even when the bandwidth is infinite, the capacity remains finite (Fig. 6.2). In the case when the requirement of the system is to minimize the energy per bit E_b rather than to maximize the spectral efficiency, then the minimum $\frac{E_b}{N_0}$ is obtained from (6.8) as;

$$\left(\frac{E_b}{N_0}\right)_{min} = \lim_{P_s \to 0}\frac{P_s}{C_{awgn}N_0} = \ln 2 = -1.59 \text{ dB} \tag{6.12}$$

This is the minimum energy for reliable communication. To achieve the minimum $\frac{E_b}{N_0}$, SNR per degree of freedom goes to zero and the delay increases. In other words for fixed bandwidth W, the communication rate goes to zero. This is the Shannon limit on the performance of any real system with AWGN transmission channel.

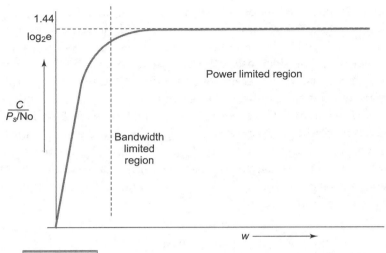

Figure 6.2   **Plot of Channel Capacity with Bandwidth.**

6.2 CAPACITY OF FADING CHANNELS

In free-space optical communication, besides the AWGN noise we have fading as well in the channel due to the turbulent atmosphere; hence the channel is considered as a fading channel. Fading channels are a class of channels with multiplicative noise instead of additive noise. The basic capacity results of the AWGN channel developed in the last section are now applied to analyze the limits to communication over these optical wireless fading channels.

Consider the baseband representation of a flat fading channel:

$$y[k] = h[k]x[k] + w[k] \tag{6.13}$$

where, $h[k]$ is the random fading process. Therefore, $|h[k]|^2$ denotes the atmospheric channel power gain. In the classical capacity formula (6.8) for the AWGN channel, $\log_2(1 + SNR)$ is the spectral efficiency [4] of the channel. As fading is a multiplicative noise, therefore,, the maximal spectral efficiency for fading channel can be expressed as:

$$\log_2 (1 + |h[k]|^2 SNR) \tag{6.14}$$

This quantity is a function of random channel gain $|h[k]|^2$, and hence the spectral efficiency is also random in this case. Here, $|h[k]|^2 SNR$ will now denote the instantaneous received SNR per symbol. As the coherence time of typical optical wireless channels is of the order of 100s' of symbols, therefore, the channel varies slowly relative to the symbol rate and, hence, can be considered to constant over block of symbols.

6.2.1 Slow Fading Channel

Let us look at the characteristics of this slow fading channel. For the common case of wind-driven turbulence, although atmospheric fades are random, they are approximately constant for time intervals of the order of a millisecond or so. As the data rates are of the order of Gbits/s, a block of very large number of symbols can experience substantially similar fading conditions. The channel gain though random but remains constant for long time duration, therefore, $h[k]$ can be considered to be constant for all k. This is also called the quasi-static scenario. The channel with random constant gain h, is modeled as an AWGN channel with received signal-to-noise ratio as $[|h|^2 SNR]$. The maximum rate of reliable communication then supported by this channel is $\log(1 + |h|^2 SNR)$ bits/s/Hz which is random due to the random channel gain h.

Depending on the nature of the application and with the above conditions, different definitions of capacity are used in literature for fading channels [4]. These are discussed in the following paragraphs.

Ergodic capacity: is defined as the average maximum mutual information per unit time with an ensemble average taken over the random gains. The ergodic capacity of a random channel with an average transmit power constraint P_T, is expressed as:

$$C = \left\{ \max_{p(x):P \leq P_T} I(X;Y) \right\} \tag{6.15}$$

where $I(X;Y)$ is the mutual information between input and output bit stream with $x \in X$ and $y \in Y$. The capacity of the channel is now defined as the maximum of the mutual information between the input and the output over all statistical distributions. If each channel symbol at the transmitter is denoted by s, the average power constraint can be expressed as

$$P = E[|s|] \leq P_T \tag{6.16}$$

Using (6.15), the ergodic or mean capacity of a single channel system with a random complex channel gain h is given by

$$C = \bar{E}\{\log_2(1 + SNR.|h|^2)\} \tag{6.17}$$

where, $\bar{E}$ denotes the expectation over all channel realizations.

Therefore, the ergodic capacity can be defined as the expectation with respect to the gains of the instantaneous capacity. The ergodic assumption requires that the codeword or the symbol extent is over at least several atmospheric coherence times, which allow coding across both deep and shallow fade channel realizations. With no delay constraints, we can code over many channel realizations and achieve reliable communication at rates up to the ergodic capacity. But when delay constraints prevent averaging over deep and shallow channel realizations, and the symbol block is restricted to just one coherence time, then, strictly speaking, the channel capacity is zero. That amounts to a chance that the fading might be so severe that the instantaneous capacity is below any desired rate. The capacity of the fading channel in the strict sense is then zero, because when the channel is in deep fade, data cannot be sent at a positive rate while making the error probability as small as desired, unlike the case of AWGN channel. In this case, a more appropriate measure of capacity is the probability that the channel can support a particular desired rate.

Outage capacity: When the data rates increase on the link, the atmospheric communication channels become better described as slow fading channels, as mentioned above. Consequently, the delay constraints of the application prevent using an extended codeword and averaging over deep fades. In such situation channel realizations is not possible by ergodic capacity. This is equivalent to communication over channels where there is a finite probability that any given transmission rate will not be supported by the channel. In such case an alternative performance measure; the ε-*outage capacity* C_ε is the appropriate measure of capacity. Now suppose the transmitter encodes data at a rate R bits/s/Hz. If the channel realization h is such that $R > \log(1 + |h|^2 SNR)$, then whatever code that is used by the transmitter, the decoding error probability cannot be made arbitrarily small. The system is said to be in outage, and the outage probability is given by:

$$p_{out}(R) = \Pr\{\log(1 + |h|^2 SNR) < R\} \tag{6.18}$$

Reliable communication can be achieved whenever the channel gain is strong enough to support the desired rate R, and outage occurs otherwise. In other words, the outage capacity [2] of the channel can be thought of which allows a maximum rate of $\log(1 + |h|^2 SNR)$ bits/s/Hz of information for the fading gain of h. Therefore, reliable decoding is possible as long as this rate of channel exceeds the target rate R.

Let SNR_{hR} denote the SNR that is required to support a rate R over the fading channel of gain h. As the channel capacity is monotonically increasing with received power as shown in (6.8), for a given channel state the probability of outage can be expressed in terms of SNR as:

$$P_{out}(R) = \Pr\{SNR_h \leq SNR_{hR}\} \tag{6.19}$$

where, $SNR_h = |h|^2 SNR$, is the SNR with channel fading gain of h. This result can be expressed in terms of the complementary cumulative distribution function (CCDF) of the $|h|^2$ as:

$$P_{out}(R) = 1 - F_c(|h|^2) \tag{6.20}$$

or for the outage probability $p_{out}(R) = \varepsilon$, from (6.20) $|h|^2 = F_c^{-1}(1 - \varepsilon)$. Then, by definition, the ε-outage capacity can be expressed as:

$$C_\varepsilon = W \log_2(1 + SNR_h) = W \log_2[1 + F_c^{-1}(1 - \varepsilon)SNR] \text{ bits/sec} \tag{6.21}$$

It is clear that the atmospheric outage capacity depends on the statistical distribution of SNR_h through its CCDF $Fc(|h|^2)$.

Next, we find out that how the outage capacity is affected with different turbulent conditions of fading [4]. One would expect more impact of fading in the low SNR regime as the received signal power levels are less. At low SNR, from (6.9) the outage capacity from (6.21) reduces to,

$$C_\varepsilon \approx F_c^{-1}(1 - \varepsilon)SNR \log_2 e$$
$$\approx F_c^{-1}(1 - \varepsilon)C_{awgn} \qquad \text{bits/sec/Hz} \tag{6.22}$$

From the above, we observe that the outage capacity is only a small fraction of the AWGN capacity at low SNR for reasonable outage probabilities. At high SNR, the outage capacity can be approximated as:

$$C_\in \approx \log SNR + \log (F_c^{-1}(1 - \varepsilon)) \tag{6.23}$$

$$\approx C_{awgn} - \log\left(\frac{1}{F_c^{-1}(1 - \varepsilon)}\right) \tag{6.24}$$

Thus, the outage capacity of fading channel is reduced by a constant value with respect to the AWGN channel and the relative loss gets smaller with increasing high SNR.

6.3 CHANNEL CAPACITY OF SINGLE-INPUT SINGLE-OUTPUT ATMOSPHERIC OPTICAL CHANNEL

From the above definition of ergodic and outage capacity, next we determine them for FSO SISO system for the slow fading optical channel.

6.3.1   Ergodic Channel Capacity of SISO

(a) With Channel Side Information (CSI) Available at the Transmitter and Receiver

The capacity of the channel can be determined with the channel characteristics being known at both the transmitter and receiver end or at either of the terminal. The capacity value will be different in each case, in general. We start with the case of assuming the marginal distribution of the channel to be known at both the transmitter and receiver. This means that we can determine the turbulence conditions and can predict the parameters of the channel distribution. This assumption is feasible as the channel parameters can be effectively measured. The channel is assumed to be uncorrelated, though in practice at high bit rates the channel has certain temporal correlation and consecutive bits propagate through similar channel conditions. But, the assumption is valid because temporal correlation in practice can be overcome by means of long interleavers, and this type of correlation does not place a constraint on the capacity [5].

To keep the calculation simple, the capacity of the channel can be calculated with the OOK modulation scheme. The system model is shown in Fig. 6.3. The channel is assumed to be memoryless with i.i.d input sequence. The transmitted OOK signal, $x \in \{1,0\}$, is propagated through the optical channel of (6.13) with random channel gain expressed as $h = \eta I$, where η the effective photo-current conversion ratio of the receiver and I the turbulence induced normalized light intensity. The distribution of intensity I can be modeled depending on the turbulence condition; varying from log-normal at low turbulence to exponential in the case of high turbulence. Assuming at low turbulence, i.e., log-normal distribution in this case, the normalized light intensity is expressed as, $I = \exp(2Z)$ where Z follows the Gaussian distribution with with zero mean and variance of σ_z^2. The received signal $y \in (-\infty, \infty)$ is continuous.

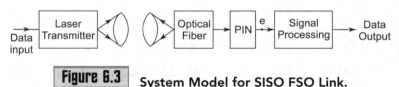

Figure 6.3   System Model for SISO FSO Link.

The channel capacity for this binary input and continuous output channel is defined as the maximum of the mutual information between X and Y over all input distributions. Also, the input distribution that maximizes the mutual information is considered to be same regardless of the channel state. The optimal input distribution for this instantaneous channel is when the probability of 1 and 0 are equal and independent of the channel state h. Hence, the ergodic capacity of this fading optical wireless channel with OOK and perfect CSI at both transmitter and receiver is given by [6] as:

$$C = \int_{-\infty}^{\infty} C_{AWGN}(h, N_0) f_h(h) dh \qquad (6.25)$$

where, $C_{AWGN}(h, N_0)$ is the capacity of the equivalent AWGN channel with binary input $\{0, h\}$ and the Gaussian noise variance of $N_0/2$. If W is the signal transmission bandwidth, then $C_{AWGN}(h, N_0)$ can be expressed as:

$$C_{AWGN}(h, N_0) = W \log_2\left(1 + \frac{h^2}{N_0}\right) \qquad (6.26)$$

In equation (6.25), $f_h(h)$ is the p.d.f. of the channel state, which is assumed to be log-normal in this case and with $f_h(h) = f_I(h/\eta)$ for η as constant can be expressed as:

$$f_I(z) = \frac{1}{2z\sigma_z\sqrt{2\pi}} \exp\left(-\frac{(\ln z)^2}{8\sigma_z^2}\right) \qquad (6.27)$$

Figure 6.4 plots the capacity curves of the log-normal fading channel for different turbulent conditions [6]. Also for comparison the plot of the AWGN non-fading channel is given. The AWGN case is the limit of the log-normal fading case when the turbulence goes to zero. As discussed in the earlier section, the channel capacity decreases with atmospheric turbulence. Also, the capacity reduces with increasing bit rates. Therefore atmospheric turbulence can be a more detrimental factor for achieving high channel throughput than low throughput.

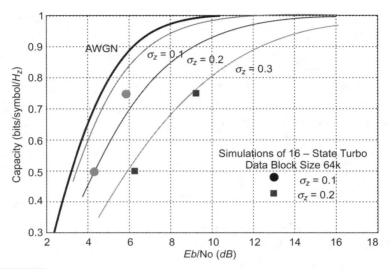

Figure 6.4   Capacity of the Log-Normal Fading Channel with OOK [6].

(b) CSI at Transmitter only

Next, let us consider the effect on capacity when CSI is only known to the transmitter. When perfect CSI is available at the transmitter, the channel capacity can be maximized to the above-determined value of the capacity when CSI is available to both sides by the best resource allocation. This is possible because once the CSI is available at the transmitter, optimal water-filling [7] resource allocation can be performed prior to transmission regardless of whether CSI is known to the receiver to achieve no loss in channel capacity. But in case, if limited CSI is available at the transmitter the capacity of the channel reduces in general.

(c) CSI at Receiver only

The ergodic capacity of the time-varying channel when the CSI is only available at the receiver can achieve the upper limit of the same level as that with CSI at the transmitter. This is possible in the case of memoryless i.i.d channel with the input distribution such that it maximizes the mutual information which is uniform and invariant of the instantaneous channel state. In this case as h is known to the decoder, the fading channel can be scaled with instantaneous power gain h^2, which is available through CSI at the receiver. The channel is then equivalent to a bandwidth-limited

AWGN channel with noise power $\dfrac{N_0 W}{2h^2}$ per dimension. The complete fading channel is then same as a time-varying AWGN channel [8].

One can summarize that the ergodic capacity of the channel remains the same in all the above cases of CSI being available at both or either side of the channel when the optimal input distribution is invariant of the instantaneous channel state [6].

6.3.2   Outage Probability of the Gaussian SISO Channel

Next we determine the outage probability for the SISO optical channel, which is defined as the percentage of time that the instantaneous rate of the channel is below a satisfactory threshold of the SNR. With the received instantaneous electrical signal power expressed as $(\eta I)^2$, mathematically the outage probability can be expressed as

$$P_{out}(\gamma^{th}) = \Pr\left(\frac{\eta^2 I^2}{N_0} < \gamma^{th}\right) = \Pr\left(I < \frac{\sqrt{\gamma^{th} N_0}}{\eta}\right) \tag{6.28}$$

where, γ^{th} is the specified threshold SNR sufficient for satisfactory reception.

Considering the case of low turbulence with log-normal p.d.f. distribution of I expressed by (6.27) [9], the cumulative distribution function of $f_I(z)$ is

$$F_I(I_{th}) = \int_{-\infty}^{I_{th}} f_I(z)dz = 1 - \frac{1}{2}erfc\left(\frac{\ln I_{th}}{2\sqrt{2}\sigma_z}\right) \tag{6.29}$$

Hence, the outage probability from equation (6.20) of this optical wireless channel will be $F_I(I_{th})$ evaluated at threshold intensity $I_{th} = \dfrac{\sqrt{\gamma^{th} N_0}}{\eta}$. Defining the average SNR of log-normal channel as, $SNR_{hL} = \dfrac{\eta^2 e^{8\sigma_z^2}}{N_0}$ and the normalized threshold SNR as $\gamma_L^{th} = \dfrac{\gamma^{th}}{SNR_{hL}}$, the outage probability is expressed as [6]:

$$P_{out}(\gamma^{th}) = 1 - \frac{1}{2}erfc\left(\frac{1}{4\sqrt{2}\sigma_z}\ln(\gamma_L^{th}) + \sqrt{2}\sigma_z\right) \tag{6.30}$$

The outage probability increases with turbulence or σ_z. The effect of turbulence is actually both ways on the performance. On one hand, the average light intensity increases with turbulence as $E[I] = e^{2\sigma_z^2}$, and therefore, the channel quality improves. But, on the other hand, with the increase of turbulence the fading increases and so the outage capacity reduces; the variance of fading increases with a much faster rate as:

$$\text{var}(I) = e^{4\sigma_z^2}(e^{4\sigma_z^2} - 1) \approx (E[I])^4 \tag{6.31}$$

It is this severe variation of the light intensity that increases the chance of outage and hence degradation of the channel. Figure 6.5 gives the outage probability of the optical channel for log-normal channel with different turbulence conditions [6].

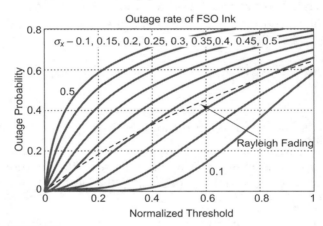

Figure 6.5   Outage Rate of Optical Wireless Channel [6].

6.4 CAPACITY OF OPTICAL FADING CHANNELS WITH DIVERSITY

We understand that diversity improves the performance of fading channels. Therefore, next, we express the capacity of the different diversity schemes for the optical channels. In the following sections, we discuss the effect of the different diversity schemes studied in the last chapter on the capacity of FSO systems.

6.4.1 Receive Diversity

As discussed in Chapter 5, the diversity of the channel can be increased by having multiple receive antennas instead of one. Multiple receive antennas can yield a N-fold power gain, and therefore, the SNR improves.

With N antennas at the receive end (Fig. 6.6a), the channel gains can be expressed by the vector:

$$h = [h_1 \ldots h_N]^t \tag{6.32}$$

The ergodic capacity of the link [10] can be calculated to be as:

$$\log (1 + \|h\|^2 \, SNR) \tag{6.33}$$

The outage will occur whenever the channel capacity is below the target rate R:

$$p_{out}^{rx}(R) = \Pr(\log\{1+ \|h\|^2 SNR\}) < R \tag{6.34}$$

In the case of high SNR, as there is a diversity gain of N, the outage probability will reduce as $\dfrac{1}{SNR^N}$. For low SNR and for small values of ε [11], the outage capacity from (6.22) is:

$$C_\varepsilon \approx F_c^{-1} (1 - \varepsilon) \, SNR \, \log_2 e$$

$$\approx (N!)^{1/N}(\varepsilon)^{1/N} SNR \, \log_2 e \ \text{bits/s/Hz} \tag{6.35}$$

With respect to the AWGN capacity, the capacity improves now by a factor of $\varepsilon^{1/N}$ with respect to no diversity.

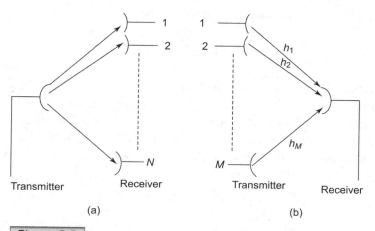

(a) (b)

Figure 6.6 (a) Receive Diversity (b) Transmit Diversity.

6.4.2 Transmit Diversity

In the case of transmit diversity, there are M transmit antennas (Fig. 6.6b) with a total power constraint of P but only one receive antenna. The capacity of the channel with channel gains $h = [h_1; \ldots; h_M]^t$ will be $\log(1 + \|h\|^2 \, SNR)$. The outage probability for a fixed rate R can be once again written as:

$$p_{out}^{tx}(R) = \Pr\{\log(1 + \|h\|^2 \, SNR) < R\} \tag{6.36}$$

which is exactly the same as the corresponding SIMO system. However, this outage performance is possible only if the transmitter has the CSI and thus can allocate more power to the stronger antennas. But when the transmitter does not know the channel gains h it does equal power transmission strategy independent of h which may not be optimum.

In the case of Alamouti scheme with $M = 2$, transmit diversity is applied without channel knowledge. The uncorrelated transmitted symbols u_1 and u_2 over a block of 2 symbol times see an equivalent scalar fading channel with gain $\|h\|$ and additive Gaussian noise with mean zero and variance of $N_0/2$. The average power in each of the two symbols is $P/2$ because of the power constraint. With channel gain of h_1 and h_2 for the two paths from the two transmitters to the receiver antenna, the capacity of the equivalent scalar channel is

$$\log\left(1 + \|h\|^2 \frac{SNR}{2}\right) \text{ bits/s/Hz} \tag{6.37}$$

The outage probability of rate R with which each of the stream, $u_1[k]$ and $u_2[k]$ is sent separately on the two antennas is:

$$p_{out}^{Ala}(R) = \Pr\left\{\log\left(1 + \|h\|^2 \frac{SNR}{2}\right)\right\} < R \tag{6.38}$$

Compared to equation (6.36), when the transmitter knows the channel, Alamouti scheme has a loss of 3 dB in the received SNR. This is because the symbols u_1 and u_2 sent from the two transmit antennas in each symbol time are independent and uncorrelated coming from two separately coded streams. Hence, the total SNR at the receive antenna at any given time instant is

$$\left(|h_1|^2 + |h_2|^2\right)\frac{SNR}{2} \tag{6.39}$$

In contrast, when the transmitter knows the channel, the symbols transmitted at the two antennas are completely correlated in such a way that the signals add up at the receive antenna and the SNR is now

$$\left(\left|h_1\right|^2 + \left|h_2\right|^2\right)SNR \tag{6.40}$$

Hence, a 3 dB power gain over the Alamouti case. The capacity performances of transmit diversity with CSI and the Alamouti scheme is plotted in Fig. 6.7 for different number of transmit antennas and with 2-PPM and 4-PPM modulation. One can see the increased capacity of MISO with CSI and some improvement in the Alamouti scheme with the PPM coding gain.

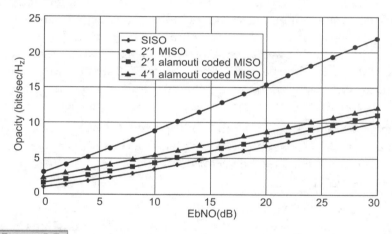

Figure 6.7   Channel Capacity of MISO with 2-PPM and 4-PPM Alamouti Scheme. For Comparison SISO and 2x1 MISO Channel with Log-Normal Distribution is Shown (σ^2=0.1).

6.4.3   Capacity of Spatially Multiplexed MIMO Channel

We take up the case of MIMO system with M transmit antennas and N receive antennas, where independent signals are sent over each transmit antenna simultaneously, which are received by all the receive antennas at the same time. The spectral efficiency in this case of spatially multiplexed MIMO channel increases linearly as compared to a logarithmic increase in systems utilizing receive diversity or no diversity. This high spectral efficiency of a MIMO system is due the fact that in a rich scattering environment of the atmosphere, the signals from each individual transmitter appear highly uncorrelated at each of the receive antennas. When the signals are conveyed through uncorrelated channels between the transmitter and receiver, the signals corresponding to each of the individual transmit antennas reach with different spatial signatures at the receiver as they pass through independent channels. The receiver can separate the signals that originated simultaneously and from different transmit antennas at the same frequency and same time instant using these differences in spatial signatures.

Similar to the case of SISO systems, in implementing MIMO systems, we must decide whether channel estimation information will be fed back or the CSI is available to the transmitter. A MIMO system with no CSI at the transmitter or without feedback is simpler to implement, and at high SNR its spectral-efficiency bound approaches that of with feedback. In most practical system, we assume that the transmitter has no channel information but CSI is available at the receiver. This is a common transmit constraint, as it may be difficult to provide the transmitter channel estimates. When CSI is not available at the transmitter, then an optimal transmission strategy is to transmit equal power from each antenna.

Figure 6.8 shows the spatially multiplexed MIMO channel with M transmit and N receive antennas. The baseband MIMO channel [10, 12] once again is modeled by the vector notation:

$$y = Hx + w \tag{6.41}$$

where, x is the $(M \times 1)$ transmit vector, y is the $(N \times 1)$ receive vector, H is the $(N \times M)$ channel matrix, and w is the $(N \times 1)$ additive white Gaussian noise vector. The channel is assumed to be flat and slow fading. Each entry of H, h_{ij} represents the path gain between the jth transmit antenna and i^{th} receive antenna. The channel for H is written as:

$$H = \begin{bmatrix} h_{11} & \cdots & h_{1M} \\ h_{21} & \cdots & h_{2M} \\ \vdots & & \vdots \\ h_{N1} & & h_{NM} \end{bmatrix}$$

In a rich scattering environment we assume the channel gains $|h_{ij}|$ to be uncorrelated and randomly distributed with distribution varying from log-normal to exponential, depending on the turbulence in the atmosphere. The capacity of a random MIMO channel with power constraint P_T can be expressed as:

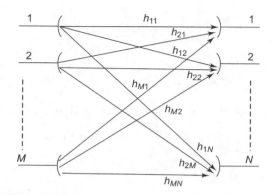

Figure 6.8 Spatially Multiplexed MIMO Channel.

$$C = E\left\{\max_{p(x):tr(\Phi)\le P_T} I(X;Y)\right\}$$

(6.42)

where, Φ is the covariance matrix of the transmit signal vector x. The total transmit power is limited to P_T, irrespective of the number of transmit antennas. By using the relationship between mutual information and entropy [2], equation (6.42) can be expanded as follows for a given H as:

$$I(x; y) = h(y) - h(y|x)$$

(6.43a)

$$= h(y) - h((Hx + w)|x)$$

(6.43b)

$$= h(y) - h(w|x)$$

(6.43c)

$$= h(y) - h(w),$$

(6.43d)

where, $h(\cdot)$ here denotes the differential entropy of a continuous random variable. It is assumed that the transmit vector x and the noise vector w are independent. As it is assumed the CSI is not available at the transmitter end hence with equal power distribution at all the transmitting antennas, the transmit covariance matrix is given by $\Phi = \dfrac{P_T}{M}I_M$, where I_M is the identity matrix. The uncorrelated noise in each receiver branch is described by the covariance matrix $\sigma^2 I_N$ with I_N as the identity matrix. The ergodic capacity for a MIMO channel can then be expressed as [13,14]:

$$C = E\left\{\log_2\left[\det\left(I_N + \frac{P_T}{\sigma^2\sigma^2 MM}HH^T\right)\right]\right\}$$

(6.44)

This can also be written as:

$$C = E\left\{\log_2\left[\det\left(I_N + \frac{SNR}{M}HH^T\right)\right]\right\} \tag{6.45}$$

where, the average SNR = $\dfrac{P_T}{\sigma^2}$ at each receiver branch and H^T is the transpose of matrix H. For M large and N fixed, by the law of large numbers, the term $\dfrac{1}{M}HH^T \to I_N$, and (6.45) reduces to:

$$C = E\{N\log_2[1 + SNR]\} \tag{6.46}$$

One can further simplify the analysis of the MIMO channel capacity by diagonalizing the product matrix HH^T either by eigenvalue decomposition or singular value decomposition in (6.44) or (6.45) [15]. Diagonalizing the product matrix HH^T can reduce the capacity expressions of the MIMO channel include unitary and diagonal matrices only. This expresses explicitly the total capacity of a MIMO channel as made up by the sum of parallel AWGN SISO subchannels. The number of these parallel subchannels is determined by the rank of the channel matrix which in general is given by:

$$\text{rank}(\mathbf{H}) = R \le \min\{M, N\} \tag{6.47}$$

The capacity will reduce for rank deficient channel matrix. This is due to the fact that a rank deficient channel matrix has some columns in the matrix, which are linearly dependent and hence the information within these columns being redundant, do not contribute to the capacity of the channel. This will be the case when the atmosphere does not transmit signals to the different receivers totally uncorrelated and independently.

By diagonalization and using (6.47) one can simplify the expression of the capacity given by (6.44) and (6.45) respectively as:

$$C = E\left\{\sum_{i=1}^{R}\log_2\left[1 + \frac{SNR}{M}\lambda_i\right]\right\} \tag{6.48a}$$

$$C = E\left\{\sum_{i=1}^{R}\log_2\left[1 + \frac{SNR}{M}\mu_i^2\right]\right\} \tag{6.48b}$$

λ_i and μ_i are the eigenvalues and singular values, respectively, of the diagonal matrices obtained by respective decomposition of the product matrix HH^T in equations (6.44) and (6.45), respectively [14]. The maximum capacity of a MIMO channel is reached in the situation when each of the M transmitted signals is received by the same set of N antennas without interference, giving a total number of $M \cdot N$ independent signals at the receiving antennas. In the situation when the signals arriving at the receivers are correlated, there are reduced number of non-zero singular values in (6.44) and (6.45), the capacity of the MIMO channel will be reduced because of a rank deficient channel matrix.

Figure 6.9 (a) gives the block diagram of a spatially multiplexed FSO system [16]. In the case of FSO MIMO frequently PPM systems are preferred over OOK. The OOK stream is not considered because of two reasons, one, bandwidth is not a constraint in MIMO and two, OOK will require dynamically changing threshold value which can be avoided in the case of using PPM. Moreover coding gain can be obtained by using Q-PPM [17, 18]. In PPM system information data is first encoded by a binary code of rate Rc. This binary encoded stream is Q-PPM modulated with rate $R = R_c \log_2 M$ (bits/channel use). The capacity of MIMO system for Q-PPM modulation for different values of M and N are shown in Fig. 6.9 (b).

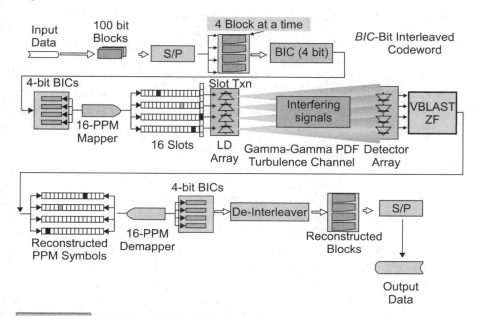

Figure 6.9 (a) System Layout Diagram of Uncoded MIMO with 16-PPM [16].

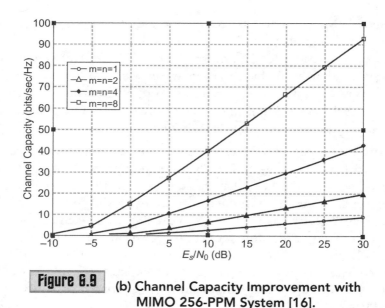

Figure 6.9 (b) Channel Capacity Improvement with MIMO 256-PPM System [16].

6.4.4 Channel Capacity of Repetitive MIMO Channels

In the previous section, we explained the linear increase in the capacity of spatially multiplexed MIMO channels when each transmit antenna sends independently coded symbols simultaneously. These signals are then received at the uncorrelated receivers, which receive independent signature of the transmitted signals at each of the receiver antenna, and therefore, can be separated by proper signal processing of the signals.

In the case of repetitive MIMO scheme, same signal is sent from all the transmitting antennas. There is a diversity gain, but not an ergodic capacity gain. The received signals at the receivers are assumed to be not correlated. At the receivers the signals can be combined with equal gain combining or select-max combining. The channel can be modelled as a quasi-static block fading channel whereby communication can take place over a finite number of blocks and each block of transmitted symbols experiencing an independent identically distributed fading. Though results for different cases of CSI being available at transmitter and receiver can be worked out, but taking the less complex and more practical situation with perfect CSI being available only at the receiver and the transmitter knowing only the channel statistics is considered.

Once again, we consider a ($M \times N$) MIMO FSO system with M transmit lasers N aperture receiver as shown in Fig. 6.8. Q-PPM modulation scheme is employed for the reasons explained earlier. Repetition transmission is employed such that the same signal is transmitted in perfect synchronism by each of the M lasers through an atmospheric turbulent channel and collected by N receive apertures assuming the lasers and apertures have negligible spatial correlation. The outage capacity of this MIMO system is calculated in the following paragraphs.

The channel is assumed to be slow. For word length of BK, with B as the number of scintillation realisations and K the time lengths, the received signal at aperture n, $n = 1, ...N$ can be written as [19]:

$$y_b^n[k] = \left(\sum_{m=1}^{M} h_b^{m,n} \right) \sqrt{\tilde{p}_b} x_b[k] + w_b^n[k] \text{ for } b = 1,...B; k = 1,......K, \quad (6.49)$$

where, $y_b^n[k], w_b^n[k]$ are the received and noise signals, respectively, for block b, time instant k and aperture n, $x_b[k]$ is the transmitted signal for block b and time instant k, and $h_b^{m,n}$ the turbulence fading coefficient of the channel between laser m and aperture n which are the independent realisations of a random fading channel H, and $\tilde{p}_b$ denotes the received electrical power of block b at each aperture in the absence of fading.

Assuming equal gain combining at the receiver, the received signal is equivalent to as obtained for a SISO channel, i.e.,

$$y_b[k] = \frac{1}{\sqrt{N}} \sum_{n=1}^{N} y_b^n[k] = \sqrt{p_b} h_b x_b[k] + w_b[k] \quad (6.50)$$

where, $w_b[k] = \frac{1}{\sqrt{N}} \sum_{n=1}^{Nr} w_b^n[k]$ are independent Gaussian random variables and h_b is defined as the normalised combined fading coefficient, i.e.,

$$h_b = \frac{A}{MN} \sum_{n=1}^{N} \sum_{m=1}^{M} h_b^{m,n} \quad (6.51)$$

where, A is a constant to normalise $E[H^2] = 1$. Thus, the total instantaneous received electrical power for block b is $p_b = \frac{M^2 N \tilde{p}_b}{A}$ and the total average received SNR is $E\left[h_b^2 p_b \right]$.

The outage probability for this PPM MIMO system for a rate above R can be obtained by:

$$p_{out}(SNR,R) = \Pr(I(p,h) < R) \quad (6.52)$$

where, $I(p,h)$ is the instantaneous input-output mutual information expressed as:

$$I(p,h) = \frac{1}{B} \sum_{b=1}^{B} I_{awgn}(p_b h_b^2) \quad (6.53)$$

where, I_{awgn} (.) is the input-output mutual information of an AWGN channel. For Q-ary PPM this is [20]:

$$I_{awgn}(SNR) = \log_2 Q - E\left[\log_2\left(1 + e^{-SNR}\sum_{q=2}^{Q} e^{\left(\sqrt{SNR}(w_q - w_1)\right)}\right)\right] \quad (6.54)$$

where, w_q is the AWGN noise signal for $q = 1,...Q$. For codewords transmitted over B blocks, it is difficult to obtain a closed form analytic expression for the outage probability though asymptotic behaviour is simpler. The simulated results are shown in Fig. 6.10 for the outage performance for different channel models [19].

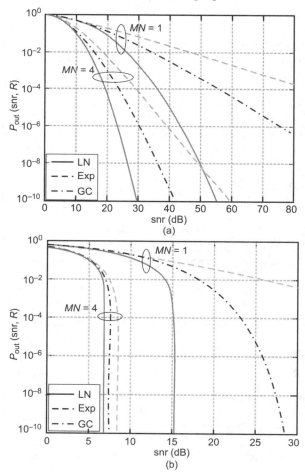

Figure 6.10   Outage Capacity of Repetitive MIMO with 2-PPM [19]; (a) with CSI at the receiver (b) CSI at the transmitter and receiver.

6.5　CAPACITY OF PHOTON POISSON CHANNELS

For optical receiver operating under the low SNR region the incident field is modeled as Poisson point process of photocurrent. This Poisson channel, like the infinite bandwidth additive Gaussian noise channel, is the only other channel for which the exact error exponent for all rates below the capacity and an explicit construction of exponentially optimal codes are known.

The shot-noise-limited operation of the receiver is modeled for an input $\lambda(t)$ which is proportional to the squared magnitude of the optical field incident on the detector at time t, integrated over its active surface. This, therefore corresponds to the number of counts registered by the direct detection device in the interval $[0, t]$, The corresponding output to this is a doubly stochastic Poisson process. This input waveform, $\lambda(t)$,. $0 \le t < \infty$, satisfies the peak and average power constraint of:

$$0 \le \lambda(t) \le A \tag{6.55a}$$

$$\frac{1}{T}\int_0^T \lambda(t)dt \le \sigma A \text{ with } 0 \le \sigma \le 1 \tag{6.55b}$$

where, the parameter A is the peak power. Also, $\lambda(t)$ defines an independent-increments Poisson counting process $v(t)$, $0 \le t < \infty$ such that:

$$v(0) = 0 \tag{6.56a}$$

and, for $\tau \ge 0$, $t < \infty$,

$$P\{v(t+\tau) - v(t) = j\} = e^{-\Lambda}\Lambda^j/j!, \qquad j = 0,1,2, \ldots \tag{6.56b}$$

where,

$$\Lambda = \int_t^{t+\tau} (\lambda(t') + \lambda_0)dt' \tag{6.56c}$$

where $\lambda_0 \ge 0$ is the background noise level or the dark current noise. Physically, the photon arrivals at the receiver are equivalent to the jumps in $v(0)$. We assume that the receiver has knowledge of $v(t)$, which it can obtain using a photon-detector.

Kabanov [21] and Davis [22] determined the channel capacity C in nats/photon for a SISO Poisson channel. This is expressed in [23] as:

$$C = A\big[p(1+s)\ln(1+s) + (1-p)s\ln s - (p+s)\ln(p+s)\big] \tag{6.57a}$$

where,　$s = \lambda_0/A$ 　　　　　　　　　　　　　　　(6.57b)

which is the ratio of noise to the signal power and where p is the 00k duty cycle given by:

$$p = \min(\sigma, p_0(s)) \tag{6.57c}$$

and

$$p_0(s) = \frac{(1+s)^{(1+s)}}{es^s} - s \tag{6.57d}$$

For the case with no dark current where, the value of p which maximizes C. and $\lambda_0 = 0$, the capacity of the Poisson channel from (6.57a), yields

$$C = Ap \ln \frac{1}{p} \tag{6.58a}$$

and $\qquad p = \min(\sigma, e^{-1}) \tag{6.58b}$

For the second case, when background noise is high, i.e. $s \to \infty$ and $p_0(s) = (1/2) + O(1/s)$, the capacity is given as:

$$C = \frac{Ap(1-p)}{2s} + O\left(\frac{1}{s^2}\right) \tag{6.59a}$$

and $\qquad p = \min(\sigma, 1/2) \tag{6.59b}$

Esentially, the quantity p gives the optimum ratio of signal energy to the maximum allowable signal energy to achieve the maximum transmission rate, i.e. $\left\{ \int_0^T \lambda(t)dt \} / AT \right\}$. Therefore, if $p_0(s) \le \sigma$, then received signals $\lambda(t)$ with $\left\{ \int_0^T \lambda(t)dt = p_0(s) \right\}$ will satisfy constraint of average power. On the other hand if $p(s) > \sigma$, then we choose signals for which $\left\{ \int_0^T \lambda(t)dt \right\} = \sigma AT$.

Thus for codes, which achieve capacity, the average number of received photons per second is pAT [23].

6.5.1   Capacity of Poisson MIMO Channel

Next the Poisson MIMO channel is considered using M transmit lasers and N direct-detection receivers. As, practically in MIMO channel the bandwidth of the atmospheric optical link is not limited by atmospheric propagation, but is rather limited by the transmitter and receiver structures, we start no channel bandwidth constraint in order to reduce the mathematical complexity.

Assuming the mth transmitter sends an intensity signal $x_m(t)$ proportional to the transmitted optical power in the interval $0 \le t \le T$ which satisfies the peak and average power constraints [24,25], respectively as following:

$$0 \le x_m(t) \le A_m \tag{6.60}$$

$$\frac{1}{T}\int_0^T E\left[x_m(t)\right]dt \le kA_m \text{ , where } 0 \le k \le 1 \tag{6.61}$$

The channel can be modeled assuming to be log-normal process during the transmission block [0, T]. The elements of the channel matrix H, give the path gain from transmitter m to receiver n as $h_{mn} = \exp(2Z_{mn})$, where Z_{mn} is a Gaussian random variable. By making the variance, $var[Z_{mn}] = -E[Z_{mn}] = \sigma_z^2$, the path gains are normalized to $E[h_{mn}] = 1$. For the shot-noise-limited direct detection photon-counting receiver, the photon count of the nth detector ($1 \le n \le N$) will be a doubly stochastic Poisson process $y_n(t)$ with rate:

$$\lambda_n(t) = \lambda_{n0} + \sum_{m=1}^{M} h_{nm} x_m(t) \tag{6.62}$$

where, $\lambda_{n0} \ge 0$ is a background noise rate. Equation (6.62) assumes all the receivers are receiving i.i.d signals from the transmitters and the signals are added incoherently at each receiver. This model also assumes that each detector observes many spatial and/or temporal modes of the background light, allowing the stochastic background noise rate to be replaced by its expected value λ_{no} [24].

The ergodic capacity of the channel is determined with the simple case of transmitter and receiver having the CSI and making the optimum use of this information to maximize the capacity. The upper bound in this case is the sum of N parallel $M \times 1$ channel capacities, given by [24]:

$$C_{UB} = \sum_{n=1}^{N} R_n I(p_n^{opt}, s_n) \tag{6.63}$$

where, $R_n = \sum_{m=1}^{M} h_{nm} A_m$ is the aggregate signal power at the nth detector. Function $I(p_n^{opt}, s_n)$ is the information function with s_n as the ratio of background noise to the signal level for the nth detector and an optimum duty cycle of $p_n^{opt} = \min(p_n^{\max}, k)$ for the capacity achieving signaling of OOK distribution , where,

$$p_n^{\max} = \frac{(1+s_n)^{(1+s_n)}}{es_n^{s_n}} - s_n \tag{6.64}$$

This is that value of p that maximizes $I(p, s_n)$. Without the average power constraint, $p_n^{\max}$ would be the optimal duty cycle of the nth MISO channel.

A lower bound [22] on the channel capacity comes because of switching between the *On* and *Off* states of the OOK signals arbitrarily fast for any time t. This once again is written as;

$$C_{LB} = \sum_{n=1}^{N} R_n I(p^{opt}, s_n) \tag{6.65}$$

where, $p^{opt} = \min (p^{max}, k)$ is the optimal OOK duty cycle in this case, and p^{max} is the value of p that maximizes $\sum_{n=1}^{N} R_n I(p, s_n)$, found by solving

$$\prod_{n=1}^{N} \left[\frac{(1+s_n)^{(1+s_n)}}{es_n^{s_n}(p^{max} + s_n)} \right]^{R_n} = 1 \tag{6.66}$$

In the low and high noise regimes the upper and lower bounds converge and one obtains a closed-form expression for the capacity of a given channel.

In the case of Poisson MIMO channel, the ergodic capacity at high signal-to-background noise ratio increases with the increase of number of transmit and receive apertures. For normalized path gains $E[h_{nm}]=1$, and with identical transmit peak power constraints it can be expressed as [24]:

$$C = NMAp^{opt} \log \frac{1}{p^{opt}} \tag{6.67}$$

This capacity can be achieved with neither transmitter nor receiver knowing the path gains.

In the case of low signal-to-background ratio regime, ergodic capacity scales as M^2 times N. Once again with condition of equal power distribution, transmit peak power constraint and with receiver background noise λ equal at all receivers, it is expressed as:

$$CN(MA)^2 p^{opt} \frac{(1-p^{opt})}{2\lambda} \tag{6.68}$$

Hence, the average capacity of the fading channel is greater than the deterministic channel with unit path gains in a noisy channel. Also, for moderate number of transmit apertures, the channel capacity does not increase appreciably by knowing the path gains at the transmitter or at the receiver.

The ergodic capacity of Q-ary PPM modulation for Rayleigh channel model are shown in Fig. 6.11 [26]. The ergodic capacity does not change with fading. But diversity and Q increases the capacity.

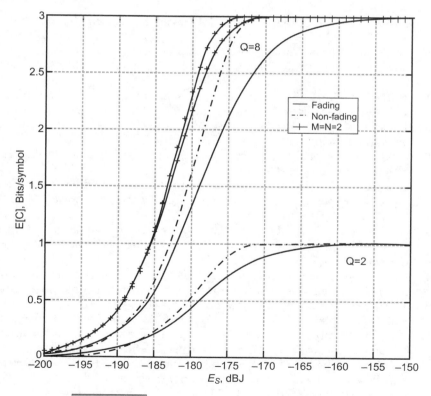

Figure 6.11 Ergodic Capacity for QPPM [26].

References

1. Shannon C.E., 'A mathematical theory of communications,' *Bell Syst. Tech. J.* 27, 379–423, 623–656 (1948).
2. Proakis I.J. and Salehi M., *Digital Communications*, Mc Graw-Hill, 2007.
3. Papoulis A., *Probability, Random Variables, and Stochastic Processes* WCB/McGraw-Hill,1991.
4. Biglieri E., Proakis J., and Shami S. (Shitz), "Fading channels: Information-theoretic and communications aspects," *IEEE Trans. Inform.Theory*, Vol. 44, no. 6, pp. 2619–2692, Oct. 1998.
5. Palomar D.P., Fonollosa J.R., and Lagunas M.A., 'Capacity results of spatially correlated frequency-selective MIMO channels in UMTS,' in 54th *IEEE Vehicular Technology Conference*, 2001. VTC 2001 (IEEE, 2001), Vol. 2, 553–557.
6. Jing Li and Uysal M., 'Optical Wireless Communication: System Model, Capacity and Coding,' *IEEE 58th Veh. Tech. Conf.*, Florida, USA, Oct. 2003, pp. 168–172.
7. Cover T.M. and Thomas J.A., *Elements of Information Theory*, Wiley-Interscience, 1991.
8. Goldsmith A.J. and Varaiya P.P., 'Capacity of fading channels with channel side information', *IEEE Trans. Inform. Theory*, Vol. 43, pp 1986–1992, Nov. 1997.
9. Zhu X. and Kahn J., 'Free space optical communication through atmospheric turbulence channels,' *IEEE Trans. Commun.* 50, 1293–1300 (2002).
10. Telatar I., 'Capacity of multi-antenna Gaussian channels,' AT&T Technical Memorandum, June 1995.
11. Theodore S. Rapparport, *'Wireless Communication: Principal and Practice'*, 2nd ed., Prentice Hall, 2002.
12. Gesbert D., Holcskei H., Gore D., Paulraj A., 'MIMO wireless channels: capacity and performance prediction,' *Global Telecommunication Conference* (GLOBECOM) 00, 2:1083–1088, 2000.
13. Foschini G.J., Gans M.J., 'On limits of wireless communications in a fading environment when using multiple antennas,' *Wireless Personal Communications*, 6:311–335, Aug. 1990.
14. Foschini G.J., 'Layered space-time architecture for wireless communication in a fading environment when using multi-element antennas', Bell Labs. *Tech. Journal*, vol. 1, No. 2, Autumn 1996, pp 41–59.
15. Shiu D., Foschini G.J., Gans M.J. and Kahn J.M., 'Fading correlation and its effect on the capacity of multi-element antenna systems', *IEEE Transaction on Communications*, Vol. 48, No. 3, 2000, pp. 502–513.
16. Sandhu H., Chadha D., 'Power and Spectral Efficient Free Space Optical Link Based on MIMO System,' *Communication Networks and Services Research Conference* (CNSR 2008), pp. 504–509, 2008

17. Djordjevic I.B., Vasic B., Neifield M.A., 'Multilevel coding in free space optical MIMO transmission with Q-ary PPM over the atmospheric turbulence channel,' *IEEE Photon. Technol. Lett.* 18, 1491–1493 (2006)

18. Wilson S.G., Brandt-Pearce M., Cao Q., and Leveque J.H., 'Free space optical MIMO transmission with Q-ary PPM', *IEEE Trans. Commun.*, Vol. 53, no. 8, pp. 1402–1412, Aug. 2005.

19. Letzepis, N. and Guillén I Fàbregas, A. '*Outage probability of the MIMO Gaussian free space optical channel with PPM*', IEEE International Symposium on Information Theory, ISIT' 08, 6–13 July 2008, Toronto, Canada.

20. Dolinear S., Divasalar D., Hamkins J., and Pollara F., 'Capacity of pulse position modulation (PPM) on Gaussian and Webb channels, JPL TMO Progress Report 42–142, Aug. 2000. http//trs-new.jpl.nasa.gov/dspace/bitsream/2014/18302/1/99-1775.pdf.

21. Kabanov Y.M., 'The capacity of a channel of the Poisson type,' *Theory Probab. Appl.*, Vol. 23, pp. 143–147, 1978.

22. Davis M.H.A., 'Capacity and cutoff rate for Poisson-type channels', *IEEE Trans. Inform. Theory*, Vol. IT-26, pp. 710–715, Nov. 1980.

23. Wyner A.D., 'Capacity and error exponent for the direct detection photon channel-Parts 1 & II', *IEEE Trans. Inform. Theory*, Vol. 34, pp. 1449–1471, Nov. 1988.

24. Haas S. and Shapiro J.H., 'Capacity of wireless optical communications,' *IEEE J. Sel. Areas Commun.* 21, 1346–1357 (2003).

25. Chakraborty K., 'Capacity of MIMO Optical Fading Channel', *IEEE International Symposium on Information Theory*, 2005, Vol., pp. 530–534.

26. Wilson S.G., Brant-Pearce M., Cao Q. and Leveque J.H., 'Free Space Optical MIMO transmission with Q-ary PPM', *IEEE Trans. Commun.*, Vol. 53, no. 8, pp. 1402–1412, Aug. 2005.

Chapter **7**

Coding in
FSO Channels

The deep fades due to turbulence in the atmosphere can cause significant errors between the input and the output data sequences of the wireless optical communication system and the probability of bit error due to these may be as high as 10^{-2}, which is not acceptable. The fundamental requirement of most wireless providers is to deliver communication links that provide uncorrupted data, voice or video with minimum delay and power requirement. For wireless optical communication links to become useful, we require FSO links with rates above 100 Mbps and error rates not exceeding 10^{-6}. There is a need, therefore, for more resilient and reliable links and efficient modulation schemes. Error control coding as well as diversity techniques then become very essential to be incorporated to improve error rate performance, and reduce the outages in the optical wireless communication system. In Chapter 5, we have discussed the diversity techniques used in FSO systems. In this chapter, we look into the different Forward Error Control (FEC) coding schemes used in FSO links for performance improvement. FEC codes can help to mitigate the effects of atmospheric turbulence-induced signal fading and increase the capacity in the intensity modulation/direct detection (IM/DD) free space optical communication links, which are normally used in the terrestrial systems.

The FEC coding involves the transmission of redundant bits in the stream of information bits in order to detect and correct a few symbol errors at the receiver. Many of the digital modulation schemes can achieve performances close to the Near-Shannon limit when implemented along with error correction codes. To improve the performance of FSO links, several FEC schemes have been proposed, including Reed Solomon codes, Turbo codes, Low Density Parity Check (LDPC) codes, etc. The atmospheric channels have few typical features; the time scale of the atmospheric fading is typically of the order of millisecs and, therefore, the optical channel has a very long memory. A fade can cause an abnormally large number of errors spanning through thousands of consecutive received channel bits. Therefore, the FEC that can be used to give good coding gain has to have a very long length. This increases the complexity of decoding and also reduces the bit rate. Instead of resorting to the use of very long and complicated codes to correct these long duration error bursts, interleaving of simpler codes, i.e., changing the order of a data sequence of the incoming symbols by permuting the data symbols also appears to be an attractive means of improving the reliability of this channel. Since the duration of the fades are random no single maximum interleaving depth can be used to render the channel completely memoryless. Furthermore, at high source data rates, interleaving depth that correspond to time separations of 1 ms between successive bits of a code word require the encoder and decoder to have a very large memory. In the case of high turbulence and high data rate simple block and tree codes are, therefore, insufficient and advanced codes such as Turbo or LDPC codes are required to be used.

In this chapter, first we give the basics of FEC coding along with certain definitions of the parameters commonly used in the different codes. Subsequently, we describe different FEC coding schemes suitable for use in FSO systems. Section 7.1 describes the basic concepts on coding. Section 7.2 gives the description of Standard Block Codes. Later in the same section, we talk of the Reed Solomon (RS) coded FSO systems. Section 7.3 describes the Tree Codes or the Convolution Codes and their implementation to FSO links. After the description of the block and convolution codes, we discuss more advanced FEC schemes – the iteratively decodable codes such as Turbo Codes and LDPC Codes in Sections 7.4 and 7.5, respectively. These are followed by their specific application in FSO system. The space time codes used in FSO were discussed earlier in Chapter 5 with the MIMO systems.

7.1 BASIC CONCEPTS

The basic objective of channel coding is to increase the resistance of the digital communication system to channel noise and fading. In channel encoding, incoming data sequence is mapped by the channel encoder to the channel input sequence. The encoded sequence is then transmitted over the channel. At the receiver, the channel output sequence is inverse mapped onto the output data sequence by the decoder. The block diagram of a coded FSO communication system is shown in Fig. 7.1. The source generates the information in the form of a sequence of discrete symbols. The channel encoder accepts the message symbols and adds redundant symbols according to a corresponding prescribed rule of the code. The encoded data is then modulated, which in turn, either internally or externally, modulates the LED/LD. The optical signals are then transmitted on the channel. On the receiver side the decoder exploits these redundant symbols to determine which message symbol was actually transmitted. The encoder and decoder consider the entire digital transmission system as a discrete channel.

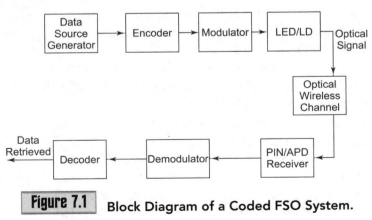

Figure 7.1 Block Diagram of a Coded FSO System.

The error correcting codes [1] can be classified in two broad categories: the *Block Codes* and *Tree Codes*. The distinguishing difference between the two codes is the presence or absence of memory element in the encoders. The encoder for Block Codes is a memoryless device, which maps a k-bit sequence of input information in n-symbol output code sequence. On the other hand the tree code encoder is a device with memory that accepts binary symbols in sets of k and outputs binary symbols in sets of n but each set of n output symbols are determined by the current input set and a span of the preceding input symbols.

Codes can be either *linear* or *non-linear*. Linear codes form a linear vector space and have the important property that the modulo-2 addition of two code words gives another codeword; in other words, the linear combination of any two code words is also a codeword.

Briefly, some of the other definitions related to coding which needs to be explained before we discuss the FEC codes are:

➤ *Hamming weight*: is the number of 1's in the codeword.

➤ *Code gain:* defined as the difference between the SNR ratio in dB per information bit needed to reach a desired bit error probability without and with coding, respectively.

➤ *Generator Matrix – G:* is a basis in a linear code which generates all the possible code words, i.e, $c_i = m_i G$, where c_i is a unique codeword of the code C, m_i is a unique row vector of information signal in the sequence M, which has been mapped with one to one correspondence.

➤ *Parity check matrix – H:* of a linear code is the generator matrix of a dual code, i.e., $cH^T = 0$.

➤ *Soft-decision and hard-decision decoding*: The optimum signal-detection scheme on an AWGN channel is the detection based on minimizing the *Euclidean distance*; the measure of the separation between two points in a signal space between the received signal and the transmitted signal. Decoding methods are classified as *hard decision and soft decision decoding*. In hard-decision decoding the output or received signal vector is first turned into a *binary sequence* $\tilde{y}$, either 1 or 0, by making decisions on individual components of the output vector, then the codeword which is closest to m in Hamming distance is chosen. On the other hand, in *soft decision*, input to the decoder is the analog or quantized equivalent of the received signals and, therefore, the error vector is also an analog or multi-valued vector. In soft-decision decoding, the *output vector* is compared with the various signal points in the constellation of the coded modulation system and the closest to it in Euclidean distance is chosen. The decoder front-end produces an integer or a *soft bit* for each bit in the data stream. This integer is a measure of how likely the bit is a 0 or 1. Thus, this introduces a probabilistic aspect to the data-stream from the front end but conveys more information about each bit than just 0 or 1.

For the conventional algebraic codes decoding is optimal and it is either maximum likelihood (ML) or maximum-a-posteriori (MAP) probability, which we have discussed in Chapter 5. Binary detection problems are frequently expressed in terms of *likelihood ratios*. For binary detection, the problem is one of determining if m_i, the transmitted message signal is 1 or 0 for the received signal y. The likelihood ratio in the case of MAP detection rule becomes as:

$LR(y) = \dfrac{P(y|m_j = 1)P(m_j = 1)}{P(y|m_j = 0)P(m_j = 0)}$ is greater or less than '1' to decide for 1

or 0, respectively.

In case of equal probability of 1's or 0's being transmitted, we obtain the likelihood ratio as:

$$LR(y) = \frac{P(y|m_j = 1)}{P(y|m_j = 0)} \tag{7.1}$$

Commonly, it is more convenient to use the *log likelihood ratio* (LLR), which is expressed as:

$$\Lambda(y) = \log \frac{P(y|m_j = 1)}{P(y|m_j = 0)} \tag{7.2}$$

The decision is made that $\hat{y} = 1$ if $\Lambda(y) > 0$ and $\hat{y} = 0$ if $\Lambda(y) < 0$.

In certain situations, there is some knowledge of the transmitted signal before decoding the received signal, e.g., some message bits are more likely to occur than others. This is the *a-priori* probability which can be used with advantage in the decoding process. On the other hand, *a-posteriori* probability refers to the information gained after decoding.

7.2 LINEAR BLOCK CODES

A linear block code involves grouping of q-ary symbols into blocks of k bits and adding (n-k) check or parity bits to form an n symbol code. There is no memory from one block to another block.

The characteristic of this code can be summarized as following:

1. Binary (n, k) Block Code is a collection of 2^k binary sequence, with code words c_i, for $1 \le i \le 2^k$. Each c_i is a sequence of length n with components equal to 0 and 1. The collection of the code words is called a *codebook* or the *code*.

2. The code rate, R is defined as k/n and the *bandwidth expansion ratio* as the inverse of the code rate. Practical values of k range from 3 to several hundred and R from 1/4 to 7/8.

3. If input sequences m_1 and m_2 are mapped to c_1 and c_2 code words, respectively, then $m_1 \oplus m_2$ is mapped to $c_1 \oplus c_2$.

4. The number of places the two codewords differ or the *digital difference* between any two code vectors c_i and c_j is the Hamming *distance*, expressed as $d(c_i, c_j)$.

   If the Hamming weight of the code vectors, i.e., the number of '1's in the codeword, be denoted as $w(c_1)$ and $w(c_2)$, then $d(c_1, c_2) = w (c_1 \oplus c_2)$.

5. The minimum distance of a Linear Block Code equals the smallest nonzero vector weight or

$$d_{min} = [w(c_1)]_{min} \quad c_1 \ne (0, 0, \dots .0) \tag{7.3}$$

6. The error correcting capability of the code is $t \left(\le \dfrac{d_{min} - 1}{2} \right)$ errors.

Encoding: of linear block code is done with the generator matrix. The generator matrix encodes a message vector of length k to a vector of length n. A *Systematic* Block Code has the property that it takes the form, in which the code word vector corresponding to each information sequence starts with the message sequence followed by some extra bits, expressed as

$$C = (m_1 m_2 \dots m_k\, x_1 x_2 \dots x_q) \quad \text{where } q = n - k \tag{7.4}$$

$$= (M|X)$$

Here, M is a k-bit message vector and X a q-bit parity *check* vector. These check digits are represented in matrix form known as *parity check matrix*. This matrix represents a set of linear homogeneous equations, and the set of solutions of these equations are the set of code words.

For a systematic linear (n,k) block code of a message vector M, the corresponding code vector C can be obtained by a matrix multiplication;

$$C = MG \tag{7.5}$$

The $k \times n$ generator matrix G, is defined as:

$$G = [I_k | P], \tag{7.6}$$

where, I_k is the $k \times k$ identity matrix and P is a $k \times q$ submatrix of binary digits. The identity matrix I_k, reproduces the message vector for the first k elements of C and the submatrix P, generates the parity check vector X as:

$$X = MP \tag{7.7}$$

Decoding: of Linear Block Codes for large k can be done by *Syndrome decoding*. For any systematic code there is a $q \times n$ parity check matrix H, defined as $H^T = \begin{bmatrix} P \\ \overline{I_q} \end{bmatrix}$, where H^T denotes the transpose of H. H^T is the generator matrix of the dual/orthogonal code C^T and I_q is $q \times q$ identity. Therefore, given H^T and a received vector Y, error detection can be based on

$$S = YH^T \tag{7.8}$$

where, S is *syndrome* of Y, which is a q-bit vector. If all elements of S equal zero, then either Y equals the transmitted code vector and there is no transmission error, or Y equals some other code vector and the transmission errors are undetectable.

Linear Block Codes can be prohibitively long for good performance in a noisy channel. Therefore, most of communication systems using Block Codes invariably use *Cyclic Codes,* a subclass of Linear Block Codes, which as the name indicates has a cyclic structure. Cyclic codes make encoding and decoding more efficient.

Encoding: A (n, k) Block Code C, *is* said to be *cyclic* if it is linear and if for every codeword $(c_0, c_1, \ldots c_{n-1})$ in C, its right shift $C' = (c_{n-1}, c_0, c_1, \ldots c_{n-2})$ is also in C.

It is easier to represent each code word as a polynomial, called the codeword polynomial $C(p)$ as:

$$C(p) = \sum_{i=1}^{n} c_{n-i} p^{n-i} = c_{n-1} p^{n-1} + c_{n-2} p^{n-2} + \cdots + c_1 p + c_0 \tag{7.9}$$

The generator polynomial is then of the form:

$$G(p) = p^q + g_{p-1} p^{q-1} + \cdots + g_1 p + 1 \tag{7.10}$$

As all the codeword polynomials are multiples of $G(p)$ and hence, similar to (7.5) can correspond to the polynomial product as

$$C(p) = M(p)G(p) \tag{7.11}$$

where, $M(p)$ is the message sequence polynomial representing a block of k message bits.

Decoding: The syndrome decoding for cyclic codes is similar at the receiver as that for the Linear Block Code. Given the received signal vector Y, the syndrome is determined from:

$$S(p) = rem\left[\frac{Y(p)}{G(p)}\right] \tag{7.12}$$

So if $Y(p)$ is a valid code polynomial then $G(p)$ will be a factor of $Y(p)$, and $[Y(p)/G(p)]$ has a zero remainder, otherwise errors have occurred.

Next, we go on to discuss the specific cyclic codes, the Reed-Solomon Codes or the RS codes, which are used in FSO systems. RS codes belong to the subclass of the Bose-Chaudhuri Hocquenghem (BCH) cyclic code family. The RS codes are considered as burst error-correcting codes and, therefore, are suitable for high-speed FSO system. They can also be used as inner encoder-decoder pairs. The coding of RS codes is based on groups of bits or bytes rather than on individual bits. This feature makes them correct burst of errors. These are non-binary codes, i.e., in a code word $c = (c_0, c_1 \ldots c_{n-1})$, the elements c_i, $0 \leq i \leq n - 1$, are members of an m-ary alphabet, where $m = 2^k$. The k-information bits are mapped into a single element of the m-ary alphabet. These m-ary symbols are then mapped into n m-ary symbols and transmitted over the channel. The encoded symbols of the code polynomial are viewed as the coefficients of an output codeword polynomial $C(p)$ constructed by the multiplication of the message polynomial $M(p)$ of maximum degree $(k-1)$ and generator polynomial $G(p)$ of degree $(n-k)$. The transmitter sends the $(n-1)$ coefficients of $C(p)$, and the receiver can use polynomial division by $G(p)$ of the received polynomial to determine whether the message is in error by determining the non-zero remainder. If $R(p)$ is the non-zero remainder polynomial, then the receiver can evaluate $R(p)$ at the roots of $G(p)$ and build a system of equations that eliminates $Y(p)$ and identifies which coefficients of $R(p)$ are in error. If the system of equations can be solved, then the receiver knows how to modify this $R(p)$ to get the most likely $C(p)$. The generator polynomial $G(p)$ for a t error correcting code is defined as:

$$G(p) = (p - \alpha^1)(p - \alpha^2)...(p - \alpha^{2t-1})(p - \alpha^{2t}) \qquad (7.13)$$

where, α, α^2, ..., α^{2t} are its roots. Hence, the degree of the generator polynomial will always be $2t$ and the (n,k) RS code, therefore, will satisfy the relation:

$$n - k = 2t \qquad (7.14)$$

The RS code for these parameters has code word $(c_0, c_1...c_{n-1})$ if and only if α, α^2,... α^{n-k} are roots of the polynomial:

$$M(p) = c_0 + c_1 p + \cdots + c_{n-1} p^{n-1} \qquad (7.15)$$

The generator polynomial $G(p)$ is the minimal polynomial with roots α, α^2,... α^{n-k} as defined above, and the code words are exactly the polynomials that are divisible by $G(p)$.

The minimum distance for the RS code is given as:

$$d_{min} = n - k + 1 \qquad (7.16)$$

and the code rate is expressed as:

$$R = \frac{k}{n} \qquad (7.17)$$

It can correct symbols up to;

$$\frac{d_{min} - 1}{2} \qquad (7.18)$$

RS coding with power efficient digital PPM modulation schemes are commonly used in the case of FSO system because of deep fades present due to fog and turbulence. The PPM modulation also has the advantage in the case of random fades for channels with direct detection as the average signal strength does not have to be accurately known for ML detection. RS codes have been used in these systems [2-3] to further improve the sensitivity. RS codes lend naturally to PPM signaling scheme, because the alphabet size can be easily matched to the PPM order. A (n,k) RS code is conventionally tailored to fit an Q-PPM modulation by choosing RS code symbols from GF(Q) and using $n = Q{-}1$, so that there is a one-to-one correspondence between PPM symbol errors and code symbol error. RS codes are optimal hard decision codes with the largest minimum distance than any other code with the same rate, block length, and field order. Though soft decision decoding has better performance, but APD detection process used in FSO many times suits well to the hard decision codes.

The block diagram of the FSO system with RS coded PPM system is shown in Fig. 7.2.

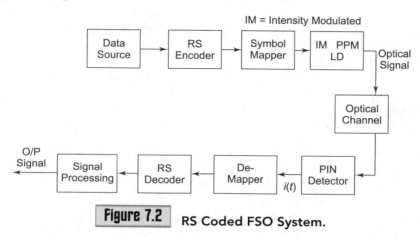

Figure 7.2 RS Coded FSO System.

The transceiver with the FSO link consists of random data source, RS encoder, symbol mapper, PPM modulator, channel, PPM demodulator, de-mapper and RS decoder. The output bits of the RS coder form sequence of symbols in the symbol mapper according to the level of PPM modulation. The binary data is first encoded into words; each word is considered as a sequence of symbols with each symbol to be $\log_2 Q$ bits long. The entire block encoded word is then transmitted as a sequence of n PPM frames. On the receiver side, ML decoding is based on the observed slot counts or received voltage level over the entire sequence. The hard decision decoding scheme is based on a PPM decision made in each frame, and the sequence of n frame decisions used in the word decoding. Figure 7.3 shows the simulated results for performance of RS (255, 127) codes with PPM modulation under ambient light and fog conditions [2] providing up to 6 db of coding gain. By using the enhanced (255, 239) RS coding, a 6 dB sensitivity improvement is expected to be achieved with only 7% overhead, i.e., additional bandwidth expansion relative to the rate [4].

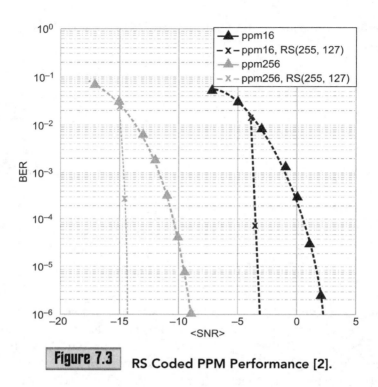

Figure 7.3   RS Coded PPM Performance [2].

Despite the performance improvements achieved by Block Codes, there are a few fundamental drawbacks to their use; the entire data code word has to be received before decoding can start, precise frame synchronization has to be achieved, the decoders for block codes work better with hard binary decisions than with soft continuous decisions and they exhibit significantly poor performance at low signal to noise ratios.

7.3   TREE CODES OR CONVOLUTION CODES

Convolution codes have a structure that effectively extends over the entire transmitted bit stream, rather then limited to codeword blocks. These codes can achieve a high performance due to the sophisticated decoding techniques used. Similar to block codes, convolution codes can correct random errors as well as burst errors. However, in contrast to block codes, convolution codes operate by adding a stream of redundant bits to a continuous flow of data bits through linear shift register. A block of k bits when mapped to a block of n bits, encoding will not only depend on the present k bit information but also on the previous information bits. In general, there are kL stages of shift register and n linear algebraic function generators that produce n output bits for every k information bits. At each instant of time, k- information bits enter the 1st stage of the shift register and the contents of the last k stages of the shift register are dropped. After the k bits have entered the shift register, n-linear combinations of the contents of the shift register are computed and used to generate the encoded waveform. Hence, the n block length output of the encoder not only depend on the most recent k bits but also on the earlier $k(L-1)$ contents of the first $k(L-1)$ stages of the shift register before the k bits arrived. As there are n output bits for every k input bits, the code rate is defined as $R = k/n$. The parameter L is called as the *constraint length* of the code. The values of k and n range typically from 1 to 8, R in the range of 1/4 to 7/8, and L in the range from 2 to 50. A convolution code does not subdivide into code words, hence the weight $w(C)$, of the entire transmission sequence generated by some message is considered. The free distance of a convolution code is defined as, $d_{free} = [w(C)]_{min}$, where C is not a zero code vector. Convolution code can be represented analytically by the *transfer function matrix* and graphically by *state transition diagram* and *the trellis diagram*. The trellis diagram is compact and most popular representation.

Encoding: We explain the encoding for the convolution codes with the help of an example shown in Fig. 7.4. The encoder in this case has $k =1$, $n =2$ and $L =2$ and rate 1/2. To initialize, all the stages of the shift register, which are S_1 and S_2 in this case, are loaded with 0 bits.

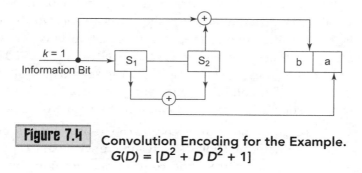

Figure 7.4   **Convolution Encoding for the Example.**
$$G(D) = [D^2 + D \; D^2 + 1]$$

In general, the data bits enter the encoder as a block of k bits at a time and the corresponding n output bits are transmitted over the channel. This continues till the last Data Block enters and then to return the encoder back to the 0 state another set of kL (in this example 2) zeros enter the encoder and the corresponding n output blocks are transmitted over the channel. The number of states for the above encoder are $2^{kL} = 4$. Unlike the case of Block Codes, the encoder is represented by a *set* of polynomials, which are called the *generator* polynomials. This set contains kn polynomials and the largest degree of a polynomial in this set is L. The polynomial for the example considered here is $G(D) = [D^2 + D \, D^2 + 1]$, where D is the number of time units of delay of particular digit to first bit which is chosen as the time origin. The states of the shift register for the incoming input bit and the corresponding coded bits are given in Table 7.1.

Table 7.1 States of the shift register for the example of Fig. 7.4

Input message bit	Current state	Coded bits
0	00	00
1	00	01
0	01	11
1	01	10
0	10	10
1	10	11
0	11	00
1	11	00

In the case of trellis representation an encoder state is shown by nodes. For binary coding, each node in the trellis has two outgoing paths, one corresponding to the input bit 0 and the other to the input bit 1. Every code word is associated with a unique path called the *state sequence* through the trellis. The trellis is the preferred representation of the encoder behavior since the number of nodes at any level of the trellis remain fixed for any number of incoming message bits. Figure 7.5 represents the trellis for the encoder of Fig. 7.4. The solid line represents the output generated by an input 0 and a dashed line indicates the output generated by an input 1.

The encoders used in convolution coding can be of two types: one with feedback and other without feedback. A recursive systematic convolution (RSC) encoder is obtained from the conventional encoder by feeding back one of its outputs to its input. An encoder with a feedback loop generates a recursive code which has an infinite impulse response while an encoder without feedback represents a finite impulse response filter. An RSC encoder tends to produce code words that have an increased weight relative to the non-recursive encoder for a given input sequence. This results in a smaller number of codewords with low weights but with increased BER performance.

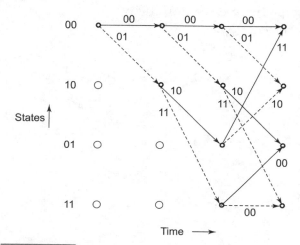

Figure 7.5 Trellis for the Example of Fig. 7.4.

Decoder: Convolution codes have the ease of incorporating soft-decision information in the decoding algorithm. There are four major classes of decoding algorithms; Viterbi algorithms, sequential decoding, threshold decoding and table look-up decoding [1]. The details of the above algorithms are beyond the scope of this book but there are many excellent literatures available with their details.

Convolution coding is one of the important FEC schemes used in FSO systems because of its better performance [5-6]. In one of the convolution interleaved coding (constraint length $L = 3$, rate $R = 1/2$) and BPPM modulation signaling schemes, studies were carried over a long clear-air weak turbulent atmospheric channel with direct detection [7]. The coding gain was directly measured as a function of interleaving delay for both hard and soft-decision Viterbi decoding. The theoretical maximum coding gains with perfect interleaving were about 7 dB for low turbulence, and 11 dB at moderate turbulence levels for the BSC channel. Practically, interleaving delays of 5 ms between the first and second channel transmission bits yielded coding gains within 1.5 dB of theoretical limits with soft-decision Viterbi decoding, and difference of 4-5 dB were observed with 100µs of interleaving delay. Soft decision

Viterbi decoding yielded 1-2 dB more coding gain than hard-decision decoding. In Fig. 7.6 gives the measured results produced for the above case [7]. Nevertheless, in the case of high turbulent atmosphere the lower bound of the pairwise error probability of convolution coded systems show a performance gap.

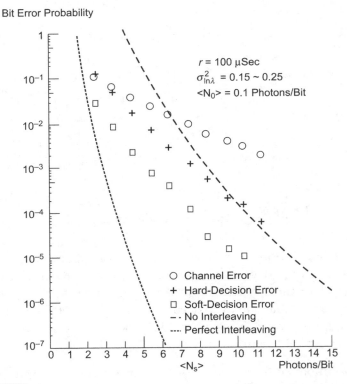

Bit Error Probability

$r = 100\ \mu Sec$
$\sigma^2_{\ln\lambda} = 0.15 \sim 0.25$
$<N_0> = 0.1$ Photons/Bit

O Channel Error
+ Hard-Decision Error
□ Soft-Decision Error
−· No Interleaving
···· Perfect Interleaving

$<N_s>$ Photons/Bit

Figure 7.6 Convolution Code Performance for Clear Air FSO System [7] With Interleaving Delay of 100 μ sec.

Both for block code and convolution code the performance can be improved by either increasing the block length of the block code or constraint length of the convolution code, respectively. They both increase the decoding complexity exponentially with n- and L, respectively. Various methods have been proposed to increase the effective length of the codes while keeping the complexity tractable. Most of these methods are based on combining simple codes to generate more complex codes. The decoding of these codes is done suboptimally, which normally performs satisfactorily. For improved performance of near Shannon coding gain advanced codes were developed, which are basically combinations of Block and Tree Codes. We next discuss in brief the *Turbo* and *LDPC* codes, which are commonly used in the outdoor optical free space links.

7.4 TURBO CODES

Turbo codes are a mix between Block and Convolution Codes. They have better performance due to the random appearance of the code on the channel, but still have a physically realizable decoding structure. Turbo codes typically have at least two convolution component encoders separated by an interleaver, and two MAP algorithm component decoders. In case of long information block and randomly selected permutations Turbo codes can achieve error rates of less than 10^{-6} at very low SNRs values around a fraction of dB. The three different arrangements of turbo codes are; Parallel Concatenated Convolution Codes, Serial Concatenated Convolution Codes and Hybrid Concatenated Convolution Codes.

Encoder: Many different types of turbo code encoders can be designed using different component encoders, input/output ratios, interleavers, and puncturing patterns. A turbo encoder for the design of parallel turbo codes is illustrated in Fig. 7.7. The encoders produce one uncoded output stream X and two encoded parity streams Y_1, Y_2 for an overall code rate of 1/3.

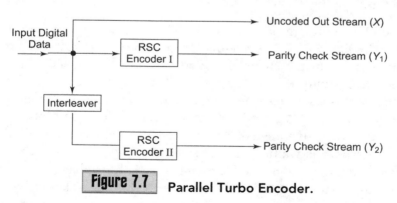

Figure 7.7 **Parallel Turbo Encoder.**

Hardware-wise, the turbo-code encoder consists of two identical systematic RSC coders connected to each other using parallel concatenation. The interleaver makes the input bits to appear in different sequences. The interleaver is a memory matrix with its length depending on the input word size. Normally, for Turbo Codes the length of these interleavers will be of the order of thousands of bits. Data is encoded by the first RSC encoder in the proper order and by the second encoder after being interleaved. The parity bits thus generated after encoding are transmitted along with the data bits as three separate data streams or sub-blocks of bits. The first subblock is the k-bit block of payload data. The second subblock is $q/2$ parity bits for the payload data, computed using a RSC code. The third subblock is $q/2$ parity bits for a known permutation of the payload data by the interleaver, again computed using an RSC code. Thus, two redundant but different subblocks of parity bits are sent with the payload. The complete block has $(q + k)$ bits of data with a code rate of $\{k/(q + k)\}$.

One of the reasons for the better performance of Turbo Codes is that they produce high weight code words. For example, if the input sequence is originally low weight, the systematic X and parity 1 output Y_1 may produce a low weight codeword. But, the parity 2 output Y_2 is more likely to be of higher weight codeword due to interleaver. The interleaver shuffles the input sequence in such a way that when introduced to the second encoder, it is more likely to produce a high weight codeword.

Decoding: For efficient data recovery, theoretically the performance analysis of Turbo codes assumes ML decoding, but as ML decoder is complex to implement in practice, it is not used. Turbo decoding is done iteratively and the performance depends upon the particular algorithm used for decoding.

The block diagram of a Turbo Decoder, which has two decoders is shown in Fig. 7.8. The decoding process starts by receiving partial information from the channels X and Y_1 and then passed to the first decoder. The information Y_2 goes to the second decoder. The first decoder yields a soft decision, and its output can be used as the a-priori information by the 2nd decoder. Each decoder produces soft-bit decisions in order to take advantage of this iterative decoding scheme. While the second decoder is waiting, the first decoder makes an estimate of the transmitted information, interleaves it to match the format of the parity 2, and sends it to the second decoder. The second decoder takes information from the channel and the first decoder and re-estimates the information. The second estimation by the second decoder is looped back to the first decoder, where the process starts again. The cycle will continue until certain conditions are met, such as, certain number of iterations to be performed. When the decoder is ready, the estimated information is finally sent out of the cycle and hard decisions are made. The result is the decoded information sequence. Multiple iterations of decoding are thus carried out to achieve considerable performance gain.

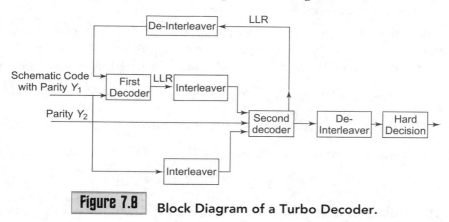

Figure 7.8   **Block Diagram of a Turbo Decoder.**

Several serial input serial output (SISO) decoding algorithms have been proposed in the literature. Though Viterbi Algorithm (VA) is widely used in the decoding of convolution codes and is an optimal method for minimizing the probability of symbol error, but in the case of Turbo Codes MAP algorithm is more often considered better, though it is computationally more intensive. A simplified version of MAP called Max-Log-MAP, which achieves a significant complexity reduction with only small performance degradation has been proposed in [8]. A modification to the Max-Log-MAP algorithm, the Log-MAP algorithm provides nearly optimum performance while still maintaining the low complexity. Also, a modified VA algorithm called the Soft Output Viterbi Algorithm (SOVA) is used to generate the soft reliability information [9]. BER performance of the MAP algorithm is superior to the VA. The iterative MAP turbo-decoding procedure though does not guarantee convergence to the ML codeword but since the turbo decoder attempts to minimize the bit-error rate rather than the word-error rate, it can in some circumstances yield a slightly lower bit-error rate than a ML decoder.

For the cases, when obtaining sufficient data from iterative simulations is impractical then it can be useful to have theoretical performance bounds for the Turbo Codes detection, as is also done for Convolution Codes. The bounds are useful in estimating the error floor which is rather difficult to measure by simulation. Transfer function bounding techniques are used to obtain the upper bounds on the BER for ML decoding of Turbo Codes [10-11]. The error probability is upper bounded by a union bound that sums contributions from error paths of different encoded weights.

In the case of turbo codes, these bounds require a term-by-term joint enumerator for all possible combinations of input weights and output weights of error events.

In order to obtain the transfer function to analyze the performance of turbo coded system the input-output weight enumerator function has to be calculated. The random interleaver of length N, maps a given word of weight d into all its $\begin{pmatrix} N \\ d \end{pmatrix}$ distinct permutations, each with equal probability of $1 \Big/ \begin{pmatrix} N \\ d \end{pmatrix}$. The transfer function which gives the input output weight enumerator function is given by [10]:

$$T(L,I,D) = \sum_{l \geq 0} \sum_{i \geq 0} \sum_{d \geq 0} L^l I^i D^d t(l,i,d) \tag{7.19}$$

where, the enumerator variable $t(l, i, d)$ denotes the number of paths or code fragments of length l generating sequence of weight i, and Hamming weight d.

For the case of M-ary orthogonal signal, one can express the conditional probability of producing codeword fragments of weight d for a given randomly selected input sequence of weight i, as:

$$p(d|i) = \frac{t^{\log_2 M}(N,i,d)}{\begin{pmatrix} N \\ i \end{pmatrix}} \tag{7.20}$$

If we choose a random selection of permutations, the probability that an input sequence of weight i will be mapped into codeword fragments of weights d_0, d_1, d_2 is given by:

$$\tilde{p}(d_0,d_1,d_2|i) = p(d_0|i)p(d_1|i)p(d_2|i) \tag{7.21}$$

The conditional probability that a ML decoder will select a particular codeword of total weight $d = \dfrac{N}{\log_2 M} - d_0 + d_1 + d_2$, to the all-zero codeword is $Q(\sqrt{2dE_s/N_0})$, where, $Q(.)$ is the complementary unit variance Gaussian distribution function. Thus, the codeword error probability P_w is upper bounded as follows:

$$P_w = \sum_{i=1}^{N} \text{Prob [error event of weight i]} \leq \sum_{i=1}^{N} \begin{pmatrix} N \\ i \end{pmatrix} E_{d|i} \left\{ Q\left(\sqrt{\frac{2dE_s}{N_0}} \right) \right\} \tag{7.22}$$

where, the conditional expectation $E_{d|i}\{.\}$ is over the probability distribution $\tilde{p}(d_0,d_1,d_2|i)$. Similarly, the information bit-error probability

P_b is upper bounded by

$$P_b = \sum_{i=1}^{N} \frac{i}{N} \text{ Prob[error event of weight i]} \leq \sum_{i=1}^{N} \frac{i}{N} \binom{N}{i} E_{d|i} \left\{ Q\left(\sqrt{\frac{2dE_s}{N_0}} \right) \right\}$$

(7.23)

Turbo coding is increasingly used in high turbulent FSO system. With increased interleaver length it can achieve good BER performance through high turbulent atmospheric fading channels, though at the cost of system complexity and delays in coding and decoding. Turbo-Coded atmospheric optical communication systems have been studied for different modulation schemes, such as OOK, BPPM and sub-carrier BPSK, QPSK both for SISO and MIMO systems. Instead of iterative decoding for error bound performance analysis simple analytical upper bounds or approximations as in (7.23) to the bit-error probability can be obtained. Here the choice of interleaver length is important to the performance bounds.

As an example, a system for the MIMO subcarrier schemes using Turbo coding is shown in Fig. 7.9. Input bits are first encoded by the turbo encoder and encoded bits then modulate the electrical sub-carrier. The bias signal is then added for subcarrier modulation to make the signal larger than or equal to zero. The positive electrical signal then in turn modulates

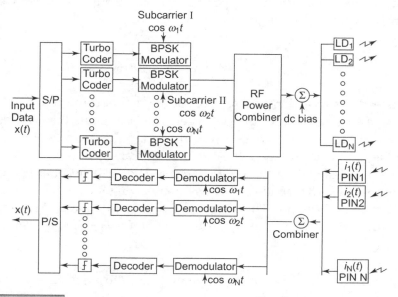

Figure 7.9 Turbo-Coded Atmospheric Optical Subcarrier System

the LD or LED to obtain the optical signal. At the receiver, the received optical signal is converted into electrical signal by the photodetectors. Then the electrical signal is demodulated and decoded, and the final data is obtained.

For the case of turbo coded atmospheric optical subcarrier BPSK system, the conditional probability that an ML detector decoder selects a particular codeword of total weight $d = d_0 + d_1 + d_2$ instead of all-zero codewords is given by [12].

$$P_w(d) = \int_{-\infty}^{\infty} \frac{1}{\sqrt{\pi}} \exp(-s^2) Q\left(\sqrt{d.SNR} \exp\left\{ \sqrt{2}\sigma_s s - \frac{\sigma_s^2}{2} \right\} \right) ds \qquad (7.24)$$

where, SNR without scintillation is expressed as

$$SNR = \frac{2(P_{max}/4)^2 (\Re G)^2 m^2}{\sigma_N^2} \qquad (7.25)$$

Here $\Re$ and G are the responsivity and average gain of the APD, respectively; P_{max} is the peak received power of the LD; m is the modulation index and, σ_N^2 is the variance of the Gaussian noise and σ_s^2 is the logarithm variance of the atmospheric scintillation due to turbulence. Using above equation the upper bound of the information bit error probability from (7.23) is given by:

$$P_b \leq \sum_{i=1}^{N} \frac{i}{N} \binom{N}{i} \sum_{d_1=0}^{N} \sum_{d_2=2}^{N} P(d_1|i)p(d_2|i)P_w(d) \qquad (7.26)$$

Instead of subcarrier BPSK/QPSK, for the case of turbo coded Q-PPM, the PPM encoder decides the slot where the laser emits the laser pulses and the PPM electrical signal simply drives LD or LED to make the optical PPM signal. At the receiver, the received optical signal is converted into the electrical signal by the photodetector. The PPM decoder compares the output of each slot and sets the symbol having the highest voltage as the transmitted symbol. The conditional probability that an ML decoder selects a particular codeword of total weight $d = d_0 + d_1 + d_2$ is given by [12]:

$$P_w(d) = \int_{-\infty}^{\infty} \frac{1}{\pi} \exp(-s^2) \int_{-\infty}^{\infty} \exp(-t^2) Q\left[\sqrt{d}SNR.\exp\{\sqrt{2}\sigma_s s - \sigma_s^2/2\} - \sqrt{2}t \right] dt ds$$

$$(7.27)$$

where, $SNR = \dfrac{(P_{\max}/4)^2 (\Re G)^2 m^2}{2\sigma_N^2}$.

Using equation (7.23) the bit error probability for atmospheric optical B-PPM can be calculated for (7.27).

Simulations results [12] for B-PPM and single subcarrier modulated BPSK are shown in Fig. 7.10 for Turbo coded performance of turbulent atmosphere. Results are also shown for convolution code FEC system as well. The upper bound on P_b becomes abruptly large for small values of E_b/N_0 as reported in literature. Thus, the turbo codes are found to be effective on the atmospheric optical channels. More powerful Turbo Codes with overhead 25% and rate 0.8 have demonstrated more than 10 dB coding gain at rates up to 10 Gbps [13].

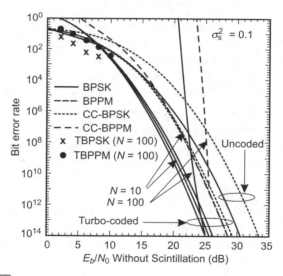

Figure 7.10   BER Performance of Turbo (1, 7/5, 7/5) Coded B-PPM and BPSK Modulation Schemes[12].

7.5   LOW DENSITY PARITY CHECK CODES

The Low Density Parity Check (LDPC) codes are considered to be the strongest among the present error control codes. These are a class of Linear Block Codes. First introduced by Gallagher in 1960 [14], but it was only in 1990 that they got the attention. As their name suggests, LDPC Block Codes have the parity-check matrices that contain only a very small number of non-zero entries. It is the sparseness of H which guarantees both, a reduced decoding complexity and a minimum distance. LDPC codes are designed by constructing a sparse parity-check matrix first and then afterwards determining a generator matrix for the code. The major difference between LDPC codes and Classical Block Codes is how they are decoded. Classical Block Codes are generally decoded with ML like decoding algorithms and so are usually short and designed algebraically to make this task less complex. LDPC codes, however, are decoded iteratively using a graphical representation of their parity-check matrix and so are designed with the properties of H as a focus. The decoder complexity of LDPC codes is significantly lower than that of serial/parallel concatenated Turbo Codes and is, therefore, a more attractive choice in optical free space transmission system of high data rates in Gbps.

A (n, k) LDPC code has a parity-check matrix with dimension of $(n–k)$ $\times n$. For the parity-check matrix with the number of 1's in each row given as w_r and number of 1's in each column as w_c, no two columns have more than two rows with 1's in same two row locations. The density of LDPC code is, w_r/n. As in the case of Linear Block Codes, due to the parity-check matrix condition of $cH^t = 0$, there are a total of $(n–k)$ linear equations, each containing w_r of the c_i's. Also, any arbitrary pair of components c_i and c_j, $i \neq j$ appear in not more than one of the $(n–k)$ equations, as no two columns have more than two rows with 1's in same two row locations as mentioned above. For example, consider a LDPC code with the parity check matrix H given as below.

$$H = \begin{bmatrix} 1000000111 \\ 0110011000 \\ 0001010101 \\ 0101101000 \\ 1010100010 \end{bmatrix} \tag{7.28}$$

It has $w_r = 4$ and $w_c = 2$. The density of this code is, 4/10 and the parity check equations are:

$$c_1 \oplus c_8 \oplus c_9 \oplus c_{10} = 0$$

$$c_2 \oplus c_3 \oplus c_6 \oplus c_7 = 0$$

$$c_4 \oplus c_6 \oplus c_8 \oplus c_{10} = 0$$

$$c_2 \oplus c_4 \oplus c_5 \oplus c_7 = 0$$

$$c_1 \oplus c_3 \oplus c_5 \oplus c_9 = 0$$

There are total of $(n-k) = 5$ equations, each containing $w_r = 4$ variables and each c_i appearing in $w_c = 2$ of the equations and no pairs of c_i appears in more than one equation.

The feature of LDPC codes to perform close to Shannon limit of a channel exists only for large block lengths. For example, there have been simulations showing specific LDPC codes that perform within 0.04 dB of the Shannon limit at a bit error rate of 10^{-6} with a block length of 10^7. Such high performance codes are irregular codes. The large block length results also in large parity-check and generator matrices.

Besides the matrix representation, LDPC codes can also be represented graphically method. The graphical representation known as *Tanner graph* provides a complete representation of the code and also helps to describe the decoding algorithm. Tanner graphs are bipartite graphs, meaning that the nodes of the graph are separated into two distinctive sets and the edges are the lines connecting nodes of two different types. Figure 7.11 is an example for such a Tanner graph and represents the same code as the matrix in (7.28). It consists of $(n-k)$ check nodes (the number of parity bits) and n variable nodes (the number of bits in a codeword). Check node, z_i, is connected to variable node, v_j, if the element h_{ij} of H is a 1. Each variable node represents a bit of the code word. Each check node represents a parity check of the code. A closed path in a bipartite graph comprising l edges that closes back on itself is called a *cycle* of length l. The shortest cycle in the bipartite graph is called the *girth*. The girth influences the minimum distance of LDPC codes, thus affecting the decoding process. The use of large-girth LDPC codes is preferable because the large girth increases the minimum distance and thus de-correlates the extrinsic information in the decoding process.

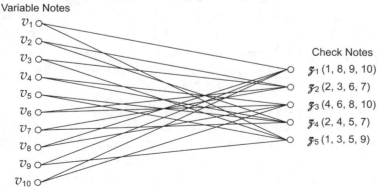

Variable Notes
v_1 v_2 v_3 v_4 v_5 v_6 v_7 v_8 v_9 v_{10}

Check Notes
z_1 (1, 8, 9, 10)
z_2 (2, 3, 6, 7)
z_3 (4, 6, 8, 10)
z_4 (2, 4, 5, 7)
z_5 (1, 3, 5, 9)

Figure 7.11   Tanner Graph Corresponding to Given Parity Check Matrix.

Regular and irregular LDPC code: In a regular LDPC code, w_c is constant for every column and $w_r = w_c$. The ratio of codeword length to the parity check constraint is also constant for every row. The example matrix from (7.28) is regular with $w_c = 2$ and $w_r = 4$. Also, from the graphical representation, there is the same number of incoming edges or degree for all the variable nodes and also for all the check nodes. But if the numbers of ones in each row or column are not constant for H, it is called an irregular LDPC code. For an irregular parity-check matrix there will be a range of weights. We designate the fraction of columns of weight i by u_i and the fraction of rows of weight i by q_i. Collectively the set u and q is called the *degree distribution* of the code. For m parity-check constraints and n codeword length, a regular LDPC code will have, $m.w_r = n.w_c$ in its parity-check matrix and for an irregular code the relationship is, $m\left(\sum_i q_i.i\right) = n\left(\sum_i u_i.i\right)$. The irregular LDPC codes can achieve reliable transmission at SNR extremely close to the Shannon limit on AWGN channel outperforming Turbo codes of the same block length and code rate.

Encoding: There are several different algorithms available to construct LDPC codes. In fact, completely randomly chosen codes are better but the encoding complexity of such codes is usually high. For large block sizes, LDPC codes are normally constructed by first studying the behavior of decoders. One way to encode the codeword is by using Gaussian Elimination method, which is a costly operation for matrices of large size as the encoding complexity grows with (n^2). Therefore, iterative algorithms are used in encoding as well as decoding which perform local calculations and pass those local results via messages iteratively. The local calculations reduce the complex processing problem to realizable levels. A sparse parity check matrix helps in several ways in this case; one, it helps to keep the local calculations simple and other, it reduces the complexity of combining the sub-problems by reducing the number of needed messages to exchange all the information.

To iteratively encode, first the message bits are placed on certain chosen variable nodes. In the second step, calculation has to be done for the missing values of the other nodes. In order to do that one needs to solve the parity check equations which involve operations with the complete parity-check matrix and the complexity would be again quadratic in the block length. In practice we use methods which ensure that encoding can be done in much shorter time by using the sparseness of the parity-check matrix. The parity-check symbols are computed by solving a system of sparse equations. For solving these equations, triangular factorizations is used to reduce the number of operations. This achieves significant reduction in complexity [15-16].

An alternative approach to simplified encoding is to design the LDPC code by algebraic or geometric methods. Such *structured codes* can be encoded with shift register circuits [17].

Decoding: The LDPC decoding algorithm comes under different names; the *Belief Propagation* (BP) [18], *Message Passing*, or the *Sum-Product* algorithm. These algorithms are iterative in nature. The reason for these names is that in each iteration, the messages are passed from variable nodes to check nodes and from check nodes back to variable nodes. Both, hard and soft decision decoding is used with BP algorithm.

With reference to Fig. 7.11, for hard decision decoding three steps are carried out iteratively. In the first step, the only information a variable node v_i, has is the corresponding received i^{th} bit y_i. All the v_i nodes send this *message* containing the bit they believe to be correct to their connected check-nodes z_j. In the second step, in turn every check node z_j calculates a response to every connected variable node. In order to do that a check node z_j looks at the message received from all the connected variable nodes and calculates the bit that the particular variable node v_i should have in order to fulfill the parity-check equation. Otherwise, this will proceed iteratively until all parity-checks are satisfied, or until the threshold number of loops is completed. Finally, in the third step the variable nodes receive the messages from the check nodes and use this additional information to decide by majority vote if their originally received bit is right. Now the variable nodes can send another message with hard decision for the correct value to the check nodes. This carries on iteratively till this corrects the transmission error and all check equations are satisfied.

Soft-decision decoding based on the BP algorithm yields a better performance and is therefore the preferred method. The underlying concept is exactly the same as in hard decision decoding except that in this case the variable node takes the soft decision. From Fig. 7.12, f_{ij} is a message sent by the variable node v_i, to the *check* node z_j. Every message contains always the pair $f_{ij}(0)$ and $f_{ij}(1)$, which stands for the amount of belief that y_i is of '0' or of a '1', respectively. Response message r_{ji} is a message sent by the check node z_j, to the variable node v_i. Again, pair

r_{ji} (0) and r_{ji} (1) indicates the amount of belief that y_i is of "0" or of "1", respectively. In the first step all variable nodes send their f_{ij} messages. Since no other information is available at this step, f_{ji} (1) $=P_i$ and f_{ji} (0) $= (1\text{-}P_i)$, where P_i is defined as:

$$P_i = P_r\,(f_i = 1|y_i) \tag{7.29}$$

In the next step the check nodes calculate their response messages r_{ji} as:

$$r_{ji}(0) = \frac{1}{2} + \frac{1}{2}\prod_{i'\in v_{p|i}}\{1 - 2f_{i'j}(1)\} \quad \text{and} \quad r_{ji}(1) = 1 - r_{ji}(0) \tag{7.30}$$

$v_{p|i}$ is for all the variable nodes except v_i.

This step and the information used to calculate the responses are illustrated in Fig. 7.12

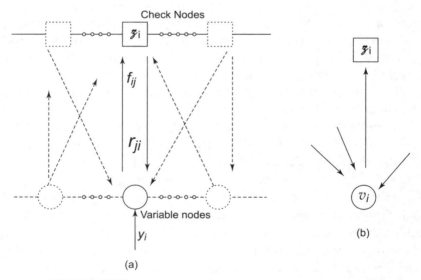

(a)

(b)

Figure 7.12   **Calculation of Belief Probabilities by Soft Decision Decoding**

The *variable* nodes then update their response messages to the check nodes. This is done according to the following equations,

$$f_{ij}(0) = A_{ij}(1 - P_i)\prod_{j'\in z_{p|j}}r_{j'i}(0) \quad \text{and} \quad f_{ij}(1) = A_{ij}P_i\prod_{j'\in z_{p|j}}r_{j'i}(1) \tag{7.31}$$

where, the constants A_{ij} are chosen to ensure that $f_{ij}(0) + f_{ij}(1) = 1$, $z_{p|j}$ includes all the check nodes except z_j. At this point, the variable nodes also update their current estimation of their variable v_i. This is done by calculating the probabilities for 0 and 1, as given below, and choosing the larger one.

$$Q_i(0) = A_i(1 - P_i)\prod r_{ji}(0) \quad \text{and} \quad Q_i(1) = A_i P_i \prod r_{ji}(1) \qquad (7.32)$$

These equations are quite similar to the ones to compute $f_{ij}(1)$ except in this case the information from every check node is used and the decision then taken:

$$\hat{v}_i = \begin{cases} 1 \text{......} if. Q_i(1) > Q_i(0) \\ 0 \text{.....................} else \end{cases} \qquad (7.33)$$

If the current estimated codeword fulfills the parity-check equations, the algorithm terminates. Otherwise termination is ensured through a maximum number of iterations.

The performance of FSO channel from low to high turbulence using LDPC codes has been evaluated and it has been found that these systems can efficiently operate across all turbulence regimes. The errors in the atmospheric FSO channel are bursty in behavior. The LDPC codes can work well with bursty channels as well, making these codes appealing for this channel. They also have large minimum distances. Some of these codes can have a regular structure and can be designed using the concepts of combinatorial design [17]. These codes also have low encoding and decoding complexity, which is a desirable feature for implementation in actual FSO communication systems. The performance improvement with LDPC codes over uncoded system and the RS codes of different rates are depicted in Fig. 7.13. The Shannon limits for the same rates are also plotted. These represent the best performance one can achieve with an infinitely long FEC code. The coding gain of the LDPC codes over an uncoded system is more than 20 dB, which is an outstanding improvement. With respect to the RS code, the LDPC at the same rate has gain more than over 5 dB at BER = 10^{-7}. The performance of the LDPC codes is less than 5 dB away from their respective Shannon limits.

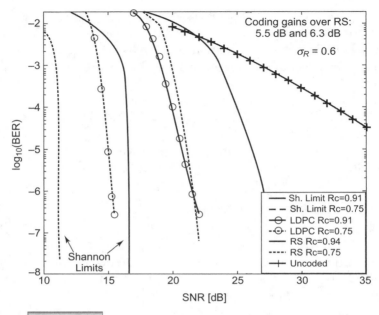

Figure 7.13   BER Performance of LDPC Codes [19].

Analyses have been done with repetitive MIMO system using component LDPC codes designed using the concept of pairwise-balanced designs [19-20]. Figure 7.14 shows the performance of LDPC codes with bit-interleaved coded modulation (BICM) or multilevel coding (MLC) schemes with 2x4 MIMO systems. These provide excellent coding gains of more than 20 dB over the uncoded system [20].

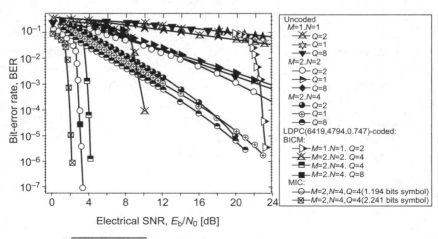

Figure 7.14   Performance of LDPC MIMO [20].

Variable rate LDPC codes also show good performance with non-repetitive MIMO antenna systems or spatial multiplexing with shallow and deep regimes of turbulence. Stronger LDPC codes are assigned in lower layer to facilitate reduction in error propagation when V-BLAST-ZF is used for the detection of the signal. The aim of implementing different parity check matrices for codewords corresponding to each layer is to avoid lower number of ones in the columns of higher rate parity check matrices, as this will render these codes unsuitable for detection by BP algorithm. The higher rate LDPC codes have been produced by drawing submatrices from the parent irregular parity check matrix by row elimination after converting it into a lower triangular form. The codewords so generated are having the same code lengths though the code rates are different. This implies that the lower layers correspond to higher rate codes to minimize the error propagated through V-BLAST detection. A simple detection technique, which is called the QR decomposition interference suppression combined with interference cancellation [21], [22], is used. The block diagram for the variable rate LDPC coded VBLAST MIMO architecture is shown in Fig. 7.15. The serial input bits are converted to parallel symbols which have to be spatially multiplexed on the different transmitting antennas. Each of the parallel streams is coded at different rates and then modulate (OOK in this case) the corresponding LED/LD. At the receiver, the received optical signal is converted into electrical signal by the photodetector. The electrical signal is detected and the LDPC decoder decodes the serial data obtained after the parallel to serial conversion.

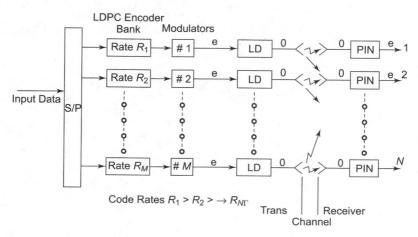

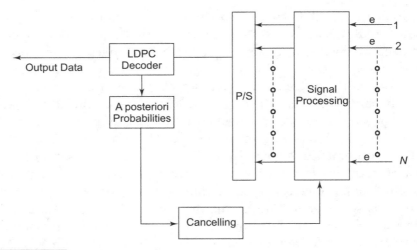

Figure 7.15   Block Diagram of Variable Rate LDPC Coded FSO System.

The simulations for (4x4) spatially multiplexed MIMO systems are shown in Fig 7.16 [23]. For the (4x4) MIMO systems, code rates used are 0.75, 0,665, 0.58 and 0.5 for transmit antenna 1, 2, 3 and 4 respectively. Comparison for two different block lengths, i.e., 200 and 1008 are shown.

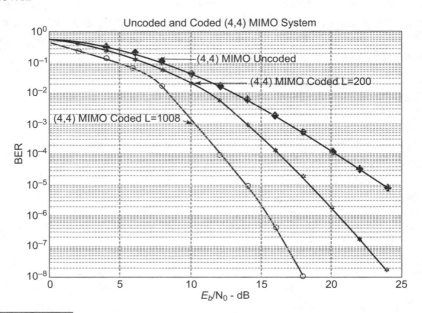

Figure 7.16   BER vs SNR for (4x4) LDPC Coded MIMO System [23].

As shown in Fig. 7.16, the (4x4) LDPC coded system provides a performance improvement of over 5dB and 9dB over the uncoded case for a block length of 200 and 1008, respectively at a BER of 10^{-6}.

Summary

Error-control codes can help to improve the performance in wireless optical communication affected by the atmospheric turbulence-induced signal fading. Though the overhead of the FEC increases the channel bandwidth required, and the block length of the code directly translates into the delay, but with the increase in the block length brings the bounds on the code rate closer to the channel capacity as opposed to the codes with smaller block lengths. Cyclic codes have a structure of polynomial rings and the length does not increase linearly as with linear block codes. All linear block codes can be made cyclic. Shift registers can be used to encode and decode cyclic codes. A Reed Solomon code is maximum distance separable code and is used for burst error correction, due to which they find application in FSO links as these have deep fades and are of longer duration. Convolution codes have memory and make decision based on past information and therefore have smaller frame lengths. Convolution codes have efficient soft decision iterative decoding algorithms available and have good performance in low and medium turbulent atmosphere. Turbo codes are a mix between block and convolution codes and use at least two convolution component encoders connected through an interleaver, and two MAP algorithm component decoders. The performance of the Turbo code is actually determined by the decoder structure. The optical wireless channel is bursty with deep and longer fades. RS codes, convolution codes and Turbo codes are good candidates for these channels. Among all the FEC codes LDPC have best performance in all the atmospheric conditions, from low to high turbulence. Exceptionally good performance is achievable with bit interleaved LDPC codes with MIMO diversity systems. For spatially multiplexed MIMO system variable rate LDPC can provide good gains. LDPC and Space-Time coding (discussed in Chapter 5) are also standards of the 3G wireless.

References

1. Todd K. Moon, *Error Correction Coding, Mathematical Methods and Algorithms,* Wiley-Interscience, 2006.

2. Sheikh Muhammad S., Javornik T., Jelovcan I., Leitgeb E., Koudelka O., 'Reed Solomon coded PPM for Terrestrial FSO Links', Proc. International Conference on Electrical Engineering (ICEE), April 2007.

3. Divsalar, D., Gagliardi, R.M., and Yuen, J.H., 'PPM performance for Reed-Solomon decoding over an optical-RF relay link', *IEEE Trans. on Communications*, Vol. COM-32, No. 3, pp. 302-305, March 1984.

4. Spellmeyer N.W., Gottschalk J.C., Caplan D.O., Stevens M.L., 'High-sensitivity 40-Gb/s RZ-DPSK with forward error correction', *IEEE Photon. Technol. Lett.* 16, 1579–1581(2004).

5. Murat Uysal, Jing (Tiffany) Li, and Meng Yu, 'Error Rate Performance Analysis of Coded Free-Space Optical Links over Gamma-Gamma Atmospheric Turbulence Channels', *IEEE Trans. Wireless Comm.*, Vol. 5, no. 6, pp. 1229-1233, June 2006.

6. Wilfried Gappmair, Markus Flohberger, 'Error Performance of Coded FSO Links in Turbulent Atmosphere Modeled by Gamma-Gamma Distributions', *IEEE Trans. Wireless Commun.*, Vol. 8, no. 5, 2209–2213, , 2009.

7. Davidson F.M., Koh Y.T., 'Interleaved convolutional coding for the turbulent atmospheric optical communication channel' *IEEE Trans on Commun.* Vol 36, no. 9, 993–1003, 1988.

8. Berrou C., Glavieux A., and Thitimajshima P., ' Near Shannon Limit Error-Correcting Coding and Decoding: Turbo-codes', *IEEE Conf. on Comm.*, Vol. 2, pp. 1064–1070, May 1993.

9. Hagenauer J., Hoeher P., 'A Viterbi algorithm with soft-decision outputs and its applications', *IEEE GLOBECOM '89*, Vol. 3, pp. 1680–1686. Nov 1989.

10. Divsalar D., Dolinar S., and Pollara F., 'Transfer function bounds on the performance of Turbo code,' TDA Prog. Rep. 42-122, Jet Prop. Lab., Pasadena, CA, pp. 4455, Aug.1995.

11. Ohtsuki T. and Kahn J.M., 'Turbo-Coded Atmospheric Optical Communication Systems', *IEEE Global Telecommunications Conference* (GLOBECOM'99), pp. 515–519, Rio de Janeiro, Brazil, Dec. 1999.

12. Tomoaki Ohtsuki, 'Turbo-Coded Atmospheric Optical Communication Systems', *IEEE, Conf. on Comm.*, Vol. 5, pp. 2938–2942, 2002.

13. Mizuochi T., et. al., 'Forward error correction based on block turbo code with 3-bit soft decision for 10-Gb/s optical communication systems', *IEEE Sel. Top. Quantum Electron.***10**, 376–386, 2004.

14. Gallager G., 'Low-Density Parity-Check Codes', IRE Transaction on Information Theory, vol. IT-8, pp. 21–28, Jan. 1962.

15. Richarson T.J. and Urbanke R.L., 'Efficient encoding of low-density parity-check codes', *IEEE Trans. Inf. Theory*, Vol. 47, no.8, pp. 638–656, 2001

16. Kaji Y., 'Encoding LDPC codes using the triangular factorization', *IEICE Trans. Fundamentals*, Vol. E89-A, No. 10, 2510–2518. Oct. 2006.

17. Kou Y., Lin S. and Fossorier M., 'Low-Density Parity-Check Codes Based on Finite Geometries: A Rediscovery and New Results', *IEEE Trans. on Information Theory*, vol. 47, no. 7, pp. 2711- 2736, Nov. 2001.

18. Jinghu Chen and Marc P.C. Fossorier , 'Near Optimum Universal Belief Propagation Based Decoding of LDPC Codes and Extension to Turbo Decoding", ISIT2001, Washington, DC, June 24–29, 2001

19. Jaime A. Anguita, Ivan B. Djordjevic, Mark A. Neifeld, and Bane V. Vasic, 'Shannon capacities and error-correction codes for optical atmospheric turbulent channels', Vol. 4, no. 9, *Journal of Opt. Net.*, pp. 586–601, Sept 2005.

20. Ivan B. Djordjevic, 'LDPC-coded MIMO optical communication over the atmospheric turbulence channel using Q-ary pulse-position modulation', 6th Aug. 2007 , Vol. 15, No. 16 , 10026–10032, *Optics Express*.

21. Namshik Kim, Hyounkuk Kim, and Hyuncheol Park, 'A variable rate LDPC coded V-BLAST system using the MMSE QR-decomposition', *IEEE Trans. Inform. Theo.*, Vol. 50, no. 11, 2824–13, 2004.

22. Jeongseok Ha, Jaehong Kim, and Steven W. McLaughlin, 'Rate-Compatible Puncturing of Low-Density Parity-Check Codes', *IEEE Trans. Information Theory*, Vol.. 52, no. 2, pp. 728–738, 2006.

23. Harinder Sandhu, Chadha D. , ' Terrestrial Free Space LDPC Coded MIMO Optical Link', Proc. WCECS 2009, Vol. 1, Oct 20–22, 2009, San Francisco, USA.

Chapter 8

FSO Link and System Design

The design of end-to-end optical wireless communication link depends on the type of wireless system under consideration, such as space/satellite, terrestrial, or an indoor system. Though the design consideration in each case is different, but in all the systems it depends on the desired link characteristics, channel constraints placed on the path and the transceiver structures used. Along with the type of the link and the distance between the sites, the design also depends on the application it is intended for; whether it is for data traffic downloads at low-speed or for uninterruptible video data or multimedia high-speed applications. The major design drivers, which affect the performance of the system, therefore, are the communication link margin allocation, data rate, bandwidth required, BER, and the size and cost of the transceiver.

In this chapter, we will look into the above issues for link planning and system design. In Section 8.1, we start with link design, which includes the considerations regarding link margin, link availability etc. In Section 8.2, the reliability of components used in FSO systems is described as it is an important consideration in the selection of the devices. Section 8.3

gives the eye safety norms used in the design of FSO systems. Section 8.4 has the details of the transceivers with the transmitter design and specifications, and similar details of the receiver section. Finally, Section 8.5, gives in brief the different noise components and how to account them in designing of the receiver.

8.1 LINK DESIGN

Figure 8.1 gives the basic block diagram of FSO terrestrial link. The FSO link design is guided by the transmission formula, which allows one to calculate the useful signal power transferred from a transmitter to a distant receiver over the desired link. The transmission formula for FSO [1] elaborates on the law of exponential decay as follows:

$$P_r = P_t \left(\frac{A_r}{A_t} \right) \eta_{TR}.K.e^{-\alpha L} \tag{8.1}$$

where, P_r is the received optical signal power, P_t is the transmitted optical power of the laser or LED, A_r is the area of the receiver aperture, A_t is the cross-sectional area of the transmitted beam at the receive telescope lens, η_{TR} is a combined transmitter-receiver optical efficiency, $K=1$ for a laser and is less than one for an LED, L is the link range, and α is the empirical attenuation coefficient. The transmitted power of the transmitter considers trade-offs between types of sources, their costs, wavelength, and permissible power levels for eye safety. The ratio of areas A_r/A_t accounts for the trade-off between beam divergence and displacement. Greater divergence means less power density and hence a weaker signal at the receiver, but also allows for more tolerance in alignment. Finally the attenuation coefficient includes all the effects of the atmospheric attenuation processes described in Chapter 2. For practical purposes α is obtained from the graphs for different atmospheric conditions plotted against wavelength.

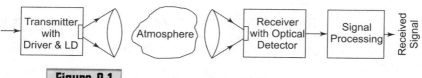

Figure 8.1 Block Diagram of FSO Terrestrial Link.

8.1.1   Link Margin

Calculation of link margin, which is obtained by detailed analysis of the power budget, is an important engineering task for any communication system. For example, in the case of link margin calculation for the optical fiber-based systems, one looks at the power coupled into the fiber from the transmitter, determines all possible losses in the fiber system to obtain the signal arriving at the receiver side. In fiber systems the losses are due to fiber, connectors, splices, and so on. The receiver has a minimum sensitivity specified at a given data rate, and the designer insures that the power coupled in the fiber at the transmitter end is sufficient to take into account the losses and still have sufficient power at the receiver to be above the minimum sensitivity for the reliable operation of the system at all times. A similar procedure has to be used in the case of FSO systems for the calculation of the link budget, which we discuss in the following paragraphs.

In this section, we derive the link margin from the transmission equation (8.1) to determine the overall system performance. For duplex operation the link consists of two transceivers one at each end. The transmitting optics shapes the transmitted laser beam which is collected by the receiver optical antenna so that the received signal is focused onto the photo detector. Consider a laser transmitter antenna with gain G_t transmitting a total power P_t at the chosen wavelength λ. The signal power received at the communication detector can then be expressed from (8.1) as:

$$P_r = P_t G_t \eta_{oTx} L_{atm} L_{free} G_r \eta_{oRx} \tag{8.2}$$

where, η_{oTx} is the transmitter optical efficiency, L_{atm} is the atmospheric loss which can be written in terms of the atmospheric attenuation factor $\alpha(dB/km) = -10 \log L_{atm}/L$, $L_{free} = \left(\dfrac{\lambda}{4\pi L} \right)^2$ is the free space loss, $G_r = \left(\dfrac{\pi D}{\lambda} \right)^2$ is the gain of the receiver antenna of diameter D, and η_{oRx} is the optical receiver efficiency. The transmitter antenna gain is $G_t \simeq \dfrac{16}{\theta_t^2}$, with θ_t as the full divergence angle at the transmitter. The received signal then can be written as:

$$P_r = P_t \left(\frac{D^2}{\theta_t^2 L^2} \right) \eta_{oTx} 10^{\left(\frac{-\alpha L}{10} \right)} \eta_{oRx} \tag{8.3}$$

While designing the link care is taken to choose the above parameters so that sufficient signal from the lasers of one transceiver reaches the photo detector on the other transceiver through the atmosphere with negligible error.

The received power as calculated in Table (8.3) can reduce with time due to long-term variation in the atmospheric characteristics as well as short-term signal fading. Also signal power reduces due to component aging, etc. In order to take care of this power variation, a certain power as link margin is accounted for while designing the link. Link margin is defined as the ratio of the available received power to the receiver power required to achieve a specified BER at a given data rate. The required power at the receiver P_{req} (watts) to achieve a given data rate B_R (bits/sec), and receiver sensitivity P_{min} (photons/bit), is related by $P_{req} = P_{min} B_R hc/\lambda$. The link margin, L_M in terms of receiver sensitivity can then be expressed as:

$$L_M = \left\{ \frac{P_t}{(P_{min} B_R hc/\lambda)} \right\} \left(\frac{D^2}{\theta_t^2 L^2} \right) \eta_{oTx} 10^{\left(\frac{-\alpha L}{10} \right)} \eta_{oRx}$$

$$(8.4)$$

The link margin can also be defined in dBm as the difference between the transmitted optical power $P_{tx}[dBm]$ by the optical source and the sum of the receiver sensitivity $P_{min}[dBm]$ combined with all the losses $L_p[dBm]$, which include the propagation loss, pointing error, building sway and so on and so forth. The link margin is then expressed as:

$$L_M = P_{tx} - Lp - P_{min} \qquad (8.5)$$

Another design parameter, which is to be considered in the case of free space communication systems performance is the *dynamic range* of the link. This is defined as the interval of acceptable power, in which the link function is guaranteed within a specified error rate. When the received power is more than the specified value the receiver is saturated, and when the received power is less than P_{min}, the required SNR or BER is not obtained.

Link Budget Computation: To give an idea of the various components of link budget, Table 8.1 shows an example of the overall calculation of the link budget of a terrestrial link. The table also shows other parameters which are used to determine link performance but not necessarily included for the link budget calculations in this example.

Table 8.1	Link budget example of a terrestrial optical wireless communication system

Parameter		Value
Wavelength (λ)	—	1550 nm
Range (L)	—	500 m
Data Rate (B_R)	—	2.5 Gbps
Receiver filter width	—	25nm
Transmitter Divergence Angle (θ_T)	—	0.1 mrad
Transmitter Antenna Gain $\left(G_T = \dfrac{16}{\theta_T^2} \right)$	—	92.04dB
Average laser power	—	30mW
Peak laser power	—	60mW
Optical background	—	0.2 W/m^2/nm/sr
BER	—	10^{-12}
Average laser transmit power	—	−15.228 dBW
Peak laser transmit power	—	−12.218 dBW
Pointing loss	—	0.5 dB
Transmitter aperture A_{tx}	—	5 cm^2
Receiver aperture A_{det}	—	15 cm^2
Geometric range loss $\left(\dfrac{\pi(L\theta_T)^2}{4(A_{det})} \right)$	—	1.17 dB
Atmospheric loss	—	20 dB/km
Atmospheric Turbulence Margin	—	7 dB
Transmitter Optical Loss	—	0.5 dB
Transmission Loss $(\lambda/4\pi L)^2$	—	192.16 dB
Receiver Antenna Gain $(\pi D/\lambda)^2$	—	98.95 dB
Receiver Optical Loss	—	0.5 dB
Link Margin	—	6 dB
Receiver sensitivity	—	−55.00 dB
Net Link Loss	—	27.84 dB
Received optical power at 2.5 Gbps	—	−43.068 dB

From the above example we find after taking into account the net link loss and link margin, the received optical power is more than the receiver sensitivity, therefore, one can expect the link to perform satisfactorily at the specified bit rate.

The above analysis for the link margin is similar to the fiber system. However, the important difference between fiber-based and FSO communication systems are related to the fact that the loss in the medium of air between the transmitter and the receiver can vary in time due to the atmospheric weather conditions. Therefore, it is very important for FSO systems to take weather conditions into consideration while designing the system.

8.1.2 Optical Link Reliability

Link *reliability* is another important issue in the FSO links. The link is said not to be reliable whenever its performance falls below the specified performance parameters for the given link range. As we have seen that line of sight connection is accomplished by means of a very narrow optical beam with low divergence, and, therefore, can get disturbed by small misalignment or atmospheric disturbance. For a successful and reliable installation of optical link, it is, therefore, necessary to know not only the steady average parameters for standard atmosphere but also the statistical characteristics of the weather at a given location. The link reliability can be quantified in terms of *availability* of the link, which is the percentage of time when the data transmission bit error rate is less or the signal SNR is more than its required value for the given range. For example, if the link availability is given as 99.99% or 99.999% for the year, then the link is said to be available with 4 nines or 5 nines reliability, respectively, over the year. In turn, the link availability can be defined in terms of link margin as the probability that any additional power losses caused either by misalignment or by atmospheric effects, including absorption and scattering are less than that of the link margin considered in the link design.

The time that the link will be available or the link availability can be expressed by means of a probability density $p(\alpha_{add})$ of the attenuation coefficient α_{add} of the additional losses as [2]:

$$T_{av} = 100\% \int_{0}^{\alpha_{add\,lim}} p(\alpha_{add})d\alpha_{add} \qquad (8.6)$$

where $\alpha_{add\,lim}$ is the limiting attenuation coefficient value given by $\alpha_{add\,lim} = L_M(L)\dfrac{1000}{L}$, for a range L. The probability distribution $P(\alpha_{add})$ can be determined by long-time monitoring of the received signal level from the measured data of the link, or using the data collected over a local area in the past for a long time. The additional loss coefficient can be obtained from visibility data [3-5] given in Chapter 2. The calculation of availability of the link at a particular site is done through the availability maps for that location. These are generated by the analyses of statistical data from weather service archives, commercial meteorology instruments, and FSO statistics combined with the system link margin [6-8].

8.1.3   Other Factors Influencing the Selection of the FSO Link

Among the many factors, FSO link selection is influenced by the following parameters:

Atmospheric Condition   Low visibility and the associated high scattering are the major limiting factors for deploying FSO systems over longer distances. Low visibility can occur during a specific time period within a year or at specific times of the day. For example, a FSO link in Delhi will experience less downtime in summer months as compared to winter in the early morning hours. In coastal areas, low visibility can be a localized phenomenon. There are few solutions to reduce the negative impact of low visibility; one is to shorten the distance between FSO terminals to maintain a specific statistical availability value. This provides a greater link margin to handle bad weather conditions such as dense fog. The second solution could be to use a multiple beam system to maintain higher link availability. Also, one could provide redundant paths to improve the availability if the visibility is limited on a local scale, for example near a water body.

Link Length   There are three ways, in which distances can affect the performance of FSO systems. First, is due to the propagation transmission loss. Second, is due to beam divergence: for a circular beam, the geometrical path loss increases by 6 dB when the distance is doubled, and lastly, is the loss due to scintillation, which accumulates with longer distances. To minimize the effects of scintillation on the transmission path, FSO systems should not be installed close to hot tar roofs or near other hot surfaces. They should be installed higher above the rooftop.

The commercially available terrestrial FSO systems have range between 0.5–5 km [9], though we know that the space FSO systems can achieve distances of thousands of kilometers.

Transmission Wavelength   Selection of wavelength depends on several factors. In general, the wavelength should not correspond to one, which is strongly absorbed in the atmosphere. Mie scattering is the dominant factor of concern for FSO terrestrial links in this regard. However, applications in dense urban areas the wavelength can also be decided by the high aerosol contents absorption. The other factors in selecting the wavelength involve factors such as availability of components, cost, required transmission distance, and so on. The preferred wavelengths are in the low attenuation 850 nm and 1550 nm wavelength band of the atmosphere. The reason for the selection of these wavelengths is also because the fiber optic systems use these bands and, therefore, a mature components technology is readily available at reasonable low cost. There is also another reason for selecting 1550 nm; the regulatory agencies allow over 50 times higher power for eye safe lasers in this band, though the lasers used cost more when compared to shorter wavelength lasers operating around 850 nm.

8.1.4 Beam Pointing and Tracking

In the outdoor links, especially the satellite links, the Pointing Aquisition and Tracking (PAT) systems have to be very elaborate and accurate because a small misalignment can cause a large pointing error. On the other hand, in the case of terrestrial FSO links relatively compact and simplified PAT systems are used. The pointing errors include both, the static pointing error with a long time constant and the random error component that has a short time constant. The static error is due to misalignment, may be due to building sway or otherwise, and the random error originates from sensor noise and control system error. Further, there are varieties of GIS mapping software programs available. These, when loaded with the PAT system, can help to achieve high resolution with the help of the 3-D topology maps with information regarding buildings and their specific locations. This helps in determine whether line of sight exists between two known locations. Figure 8.2 [9] shows one of the commercial FSO transceiver with the PAT system integrated with it.

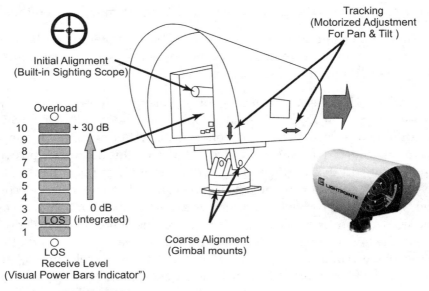

Figure 8.2   Commercial Transceiver with PAT [9].

In practice rooftop to rooftop deployments of the links are normally done for FSO terrestrial links. But, it is also feasible to place the transceivers behind windows in the building when roof access is not available. The beam should be perpendicular, or preferably making an angle around 5° to the perpendicular in order to reduce bounce back of the beam to its own receptor. Once the system is aligned, the FSO system has to maintain the alignment of the transceiver, which is a challenge because of the narrow beam of light with very little spread at the other end of the link receiver.

The alignment of the beam can get disturbed with time due to building sway and cause interruption of communication. The movement of the buildings may be due to thermal expansion, wind sway, and vibration. Building movement can usually be of either low, moderate, or of high frequency. Low-frequency motion is assigned to periods from minutes to months and is dominated by daily and seasonal temperature variations. Moderate frequency motion has periods of seconds and includes wind-induced building motion. High-frequency motion has periods of less than a second and is generally due to vibration induced by working of large machinery, movement of people, shutting doors, etc. Building sway misalignment can be avoided either by active tracking to compensate of beam dancing due to building movement, or by using a wide divergence beam such that both transmitter and receiver have large FOV. However, this gives larger loss because of decreased intensity. It has been observed that in most cases the motion of the building due to various reasons specified above has frequency content less than 10 Hz and the peak angular base motion will then not exceed 1 milli-radian for vibration of the order of 1 Hz. Hence, if the mounting hardware is well designed and installed then the misalignment due to base motion will not amplify to cause any significant error.

8.2 COMPONENT RELIABILITY

The link reliability of FSO system discussed above has two parts—one is the reliability of the channel and other that of the components used in the transceiver. For the system to be reliable, both parts have to be taken account of. Component reliability is an important factor to ensure the proper operation of service provider grade transmission equipment. The mean-time-to-failure (MTTF) values of components are especially important, as the basic requirement is to guarantee a lifetime cycle of the installed equipment well beyond its anticipated use in the network. The Telcordia qualification is of an additional value since these standards [10-12] often test for mechanical and thermal levels that are similar to many of the environmental requirements for space-based or terrestrial systems.

Active components such as laser sources, amplifiers and other electronic components have to be chosen very carefully as they are exposed to more stress factors due to their internal electronic structure. Shorter and longer wavelength Vertical Cavity Surface Emitting Lasers (VCSEL) as well as lower power Distributed Feedback Laser (DFB) lasers at longer wavelength are among the highest lifetime laser components and, therefore. are recommended for use in FSO systems. These lasers are designed to perform without significant failure for more than 10^5 hours without degradation and even under difficult operating conditions such as high wind, temperature environments, etc. For these reasons manufacturers who offer carrier-class FSO equipment generally use VCSEL in the shorter infrared wavelength range and Fabry-Perot (FP) or DFB lasers for operation in the longer IR wavelength range [13].

8.3   EYE SAFETY CONSIDERATION

High-power laser beams can cause injury to skin and eye [14], but the risks of injury to the eye are more significant because the light falling on the eye gets focused onto the retina and thereby concentrate the optical energy, hence increasing its intensity. In general, any laser that is considered to be *eye-safe* is also considered to be *skin-safe*.

The light from the wavelengths between about 400 to 1400 nm are focused by the eye onto the retina. Wavelengths above 1400 nm tend to be absorbed by the front part of the eye which is the cornea before the energy is focused and concentrated. If light is absorbed before being focused on the retina, the retina is not exposed to light and will not be damaged. The absorption coefficient of the cornea is much higher for wavelengths greater than 1400 nm than for shorter wavelengths. The visible radiation is, therefore, more likely to damage the eye than from longer wavelengths in the infrared spectrum, but there is also a natural aversion of eye to wavelengths in the visible region. On the other hand, IR being invisible to eye, wavelengths longer than 700 nm do not trigger an aversion response. Although infrared light can damage the surface of the eye, the damage threshold is higher than that for light with shorter wavelength.

The wireless optical transmitter consists of a single or a number of sources with a corresponding optics to shape the beam. LED and LD are employed as the optical radiating element but their transmission power has to be limited to eye safety regulation. The LEDs are large area devices that cannot be focused on the retina. LDs on the other hand are collimated sources whose energy can be focused on the retina. This means that a much lower launch power has be used in order to be considered eye safe.

Laser safety standards are available and are classified according to the type and power of laser. These classifications generally divides the laser power into four groups, Class 1 through Class 4, Class 1 with minimum power and Class 4 being the most powerful. Under the new International Electrotechnical Commission (IEC) standard, (IEC60825-1) [15], specific laser classes are identified, and each class is assigned specific labelling and warning instructions as shown in Table 8.2. IEC standard defines three types of locations; unrestricted, restricted, and controlled location types and installation compliance requirements based on the level of emitted power. It defines specific hazardous zones in front of the transmit aperture that must be cleared for eye-safe viewing, and restricts the installation of certain high-power laser systems in areas that are easily accessible to the public.

| Table 8.2 | IEC Standard IEC60825-2 [15] |

Hazard level	Location type		
	Unrestricted	**Restricted**	**Controlled**
1	No requirements	No requirements	No requirements
1M	Class 1 from connectors that can be opened by an end-user[1] No labelling or marking requirement[2]	No labelling or marking required if connectors that can be opened by end-user are Class 1. If output is Class 1M then labelling or marking is required 2>.	No requirements
2	Labelling or marking[2]	Labelling or marking[2]	Labelling or marking[2]
2M	a) Labelling or marking[2] and b) Class 2 from connector[1]	Labelling or marking[2]	Labelling or marking[2]
3R	Not permitted 3, 4	a) Labelling or marking[2] and b) Class 1M from connector[1]	Labelling or marking[2]
3B	Not permitted 3, 4	Not permitted 3, 4	a) Labelling or marking[2] and b) Class 1M or 2M from connector[1]
4	Not permitted 3, 4	Not permitted 3,4	Not permitted 3, 4

[1]Where the information contained in this annex differs from the requirements contained in Clause 4. The requirements of Clause 4 have precedence.

[2]Reference to Class X' in the table above means access to radiation that is within the accessible emission limits corresponding to Class X. As given m IEC 60825*1.

FSO system mostly uses Class 1 and Class 1M laser light sources. Class 2 and Class 2M cover the visible wavelength. Due to aversion response, Class 2 and Class 2M allow larger power than Class 1 and Class 1M, but no commercial FSO system uses visible laser. IEC also provides IEC60825-12 to cover FSO system classification and usage. As a matter of fact, eye-safe system has to consider not only the wavelength, but the power level as well. The new regulation considers the power density in front of the transmit aperture rather than the absolute power created by a laser diode inside the equipment. The new regulation also states that a Class 1M laser system operating at 1550 nm is allowed to transmit approximately 55 times more power than a system operating in the shorter IR wavelength range, such as 850 nm, when both have the same size aperture lens. Table 8.3 gives the IEC standard IEC60825-12 for Class 1 and 1M laser systems, both for the 850- and 1550-nm wavelengths. Class 1 lasers are safe under reasonably foreseeable operating conditions including the use of optical instruments for intra-beam viewing. Also Class 1 systems can be installed in any location without restriction.

Table 8.3   **IEC standard (IEC60825-12)**

Laser Classification	Power (mW)	Aperture Size (mm)	Distance (mm)	Power Density (mW/cm^2)
850-nm Wavelength				
Class 1	0.78	7	14	203
		50	2000	0.14
Class 1M	0.78	7	100	203
	500	7	14	1299.88
		50	2000	25.48
1550-nm Wavelength				
Class 1	10	7	14	26 00
		25	2000	204
Class 1M	10	3.5	100	10399
	500	7	14	1299.88
		25	2000	101.91

Table 8.3 shows the plot of the allowed Class 1-emitted power with beam divergence. This is shown for three types of sources, 850 nm point source, diffuse sources with different diameters at 850 nm, and for a 1500 nm point source [16]. As observed in the graph, increasing the source diameter increases the size of the image on the retina of the eye, thus reducing the possibility of thermal damage. In the case of 1550nm the light is not focused on retina, therefore, the emitted power is allowed to reach value up to 10 mW. As is understood that more divergent sources are less hazardous the eye cannot collect all the radiation. Sources of small diameter can be made diffused using a ground glass plate, or using

diffuser elements such as holographic plates, etc [17]. The effect of the diffuser is to increase the apparent size of the source, and the graph shows the benefit of this.

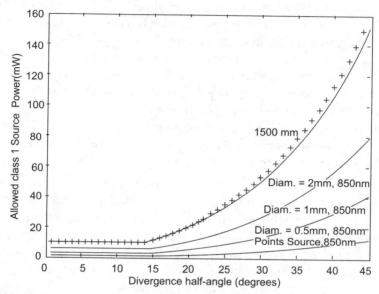

Figure 8.3 Allowed Emitted Power for Class 1 Eye-Safe Operation of Transmitter as a Function of Beam Divergence [16].

8.4 TRANSCEIVERS

In this section, we focus on design options for the transmitter and receiver blocks of the communication link shown in Fig. 8.4. The fundamental characteristics of optical sources, modulators, detectors, amplifiers, and associated noise sources will be discussed. Practical tradeoffs and implementation issues that arise from using various technologies for receiver design and optimization will be discussed.

In clear atmosphere, a well-aligned free-space communication system performance directly depends on the efficiency of optical transmitter and the sensitivity of the receiver. Improvement in receiver sensitivity can be directly used in reducing the transmitted power requirements, which in turn can lead to significant size, weight, power and cost reductions. We know that the overall transceiver system level performance is dependent on factors, such as transmitter and receiver antenna aperture sizes, wavefront quality, pointing acquisition and tracking besides the transceivers design itself. These systems are designed to operate in the windows of 780-850 nm and 1520-1600 nm in the low loss atmospheric windows, which is same as that of the fiber. As commercial-off-the-shelf modular systems with high-reliability and high performance have already been developed for the telecommunication fiber industry in these wave bands, the use of these components has helped to a great extent to the development of free-space communication systems. Around 850 nm reliable, inexpensive, high-performance transmitter and receiver components are readily available. Highly advanced VCSEL and sensitive PIN or APDs are available for operation near 850 nm. For the longer wavelength range of 1520–1600 nm, high quality transmitter and receiver components are also available. These wavelengths are better from the eye-safety point of view.

For the FSO systems for terrestrial links, both the transmitter and receiver blocks, i.e., the transceiver are developed in a single circuit board or integrated circuit chip, which provides the full-duplex communication. For example, arrays of VCSELs are made on the GaAlAs substrate and are integrated on the chip with the arrays of driver circuit electronics providing complete transmitter functionality. A similar approach is used with the receiver photodetector array and the signal detection electronics. This integration can be done at a Wafer scale, ensuring digital control circuitry to be integrated with the transmitter and receiver circuits. This not only reduces the cost, but also these optoelectronic devices can then be tested on a wafer scale [18, 19]. All the components can then be assembled into a compact package and special casing protection from rain and direct sunlight for outdoor applications is provided.

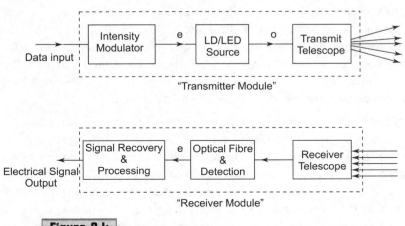

Figure 8.4   **Block Diagram of an FSO Transceiver.**

Besides the electrical and optical characteristic, the design considerations in these modules also look for small footprint and low power consumption. This is important for overall system design and maintenance. The transmitter and receiver should have the capability to operate over a wide temperature range without showing any major performance decay or degradation as they have to work through the extremes of outdoor environment, and should also have large value of MTTF operation exceeding 10 years.

In the next sections, we see the details of the transmitters and receivers.

8.4.1 Laser Transmitters

For direct detection terrestrial links, highly efficient laser transmitters capable of multi-gigabits bit rate, large bandwidth modulation at moderate average power levels, and high spatial beam quality are required. The transmitter selection is driven by divergence of the transmit beam, required quality for optics, the peak and average power, commercial availability and cost. Besides this, the factors that impact optical communication transmitter performance include the extinction ratio, output polarization state, wavelength of operation and the electrical-to-optical conversion efficiency.

The power limitation in the case of FSO transmitter is mainly because of the eye-safety consideration unlike the case of fiber-optic transmitters where the average and peak power delivered over the fiber channel is generally limited by channel nonlinearities. The FSO link has no such constraint, and hence the power limitations within the FSO transmitter are several orders of magnitude higher than those used in fiber networks. The optical transmitters are typically average power limited, which enables the average output power to be independent of the transmitted waveform shape or duty-cycle. This enables optical communication systems to use pulse shaping to improve the system performance.

Typically, the optical transmitter front end consists of a driver circuit, optical source, laser diode or LED, a Peltier element for cooling if LD is used, a modulator block, and a lens as shown in Fig. 8.5. The transmitter may consist of a single or a number of sources in the case of MIMO system, with transmitter optics or antenna to shape the beam.

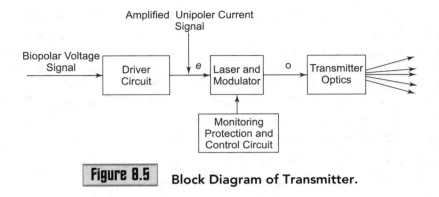

Figure 8.5 Block Diagram of Transmitter.

8.4.1.1 Sources

For optical wireless transmitter, LDs are preferred over LEDs because they have higher optical power outputs, broader modulation bandwidths and linear electrical to optical signal conversion characteristics [20]. Linearity in signal conversion is particularly important for analog and advanced modulation schemes, such as multi-subcarrier modulation or multilevel signaling. LD is linear after the threshold while LED is linear near the origin. On the other hand, LEDs are large-area emitters and thus can be operated safely at relatively higher powers. They are also less expensive. To compensate for lower powers, array of LEDs can be used. However, LEDs cannot be used beyond 100 Mbps, whereas LDs can be used for transmission at bit rates of order of 10's of Gbps with external modulation. Both LED and LD have suitable characteristics, hence, selection of one over the other depends on the application. While selecting the source the other considerations to be taken into account are of the optical energy launched by the source at angles that optimize the transmitted beam, it should be reasonably monochromatic with sufficiently high intensity and have long lifetime. Spectral emission of LD remains more stable with temperature as compared to LED but LD requires temperature stabilization. A Peltier element is used with a control circuit for this purpose, therefore, it requires more complicated electronic circuits as compared to LEDs. Nevertheless, in the outdoor environment, the properties of LDs, such as narrow spectra, high-power launch capabilities make these devices the better optical source for long distance directed LOS links. Recent developments in the VCSEL sources have made them sources of choice for both indoor and outdoor FSO application. They have well-controlled narrow beam properties, high modulation bandwidth, high speed of operation, excellent reliability and low power consumption. In the subsequent paragraphs we give certain further details of these sources.

Semiconductor Laser Sources　Present-day semiconductor laser diodes, which are used in fiber communication and also being used in FSO systems are robust, compact, efficient and have MTTF exceeding 100 years (~1×10^6 h) [21]. These lasers can be directly modulated at speeds approaching the relaxation oscillation frequency. At output power levels of 10–25 mW, the modulation bandwidth can exceed 10 GHz. This is particularly useful for subcarrier-multiplexing, where many narrow-band MHz analog signals electrically modulate a higher RF carrier which in turn modulate a single laser source. In the case of digital applications, direct modulation rates are practically limited to lower values, e.g., a few GHz, above which there is penalty due to chirping [22].

Different types of laser structures can be used in the case of terrestrial links. Fabry-Perot (FP) lasers have the simplest structure, but usually have output of multiple longitudinal lasing modes, and therefore, are not well suited as signal sources for high-performance FSO links. For stable single-frequency operation, distributed-feedback (DFB) lasers are most commonly used throughout the telecom industry along with the terrestrial FSO. The single mode operation provides a side-mode-suppression ratio exceeding 45 dB. DFB lasers based on InGaAs/InP semiconductor technology with operating wavelengths around 1550 nm have high-power output with high modulation speed, wavelength stability, reliability and long lifetimes. Higher-power DFB lasers with output powers beyond 100 mw are used in space communication only. Semiconductor lasers with integrated distributed Bragg reflectors (DBR) are mostly used in Space communication but not in terrestrial links.

Vertical Cavity Surface Emitting Laser　Development in VCSEL structures over the last decade or so has shown large improvement to make them very popular in the communication industry. There are VCSELs now in 850 nm with direct modulation speeds beyond 3 Gbps at power levels in excess of 10 mW. Also, direct electrical modulation of VCSEL lasers beyond 10 Gbps have been demonstrated and commercialized for OC-48 and 10 gigabit Ethernet operations. VCSEL lasers can operate at very low threshold currents of few milliamps and have high conversion efficiency. No active cooling of the VCSEL structure is required. In addition, VCSELs emit light in the form of a circular beam pattern making the coupling to fiber easier, and being surface emitters the coupling efficiency is much higher as compared to a standard DFB laser. But most importantly is the high lifetime with a MTTF value of more than 4×10^7 hours at 35°. At high ambient temperatures where the junction temperature can reach 90° for extended periods of time, a MTTF value of 3.9×10^5 hours or 44 years was estimated [23]

8.4.1.2 Driver Circuit

The core function of the driver is voltage to current conversion of the input voltage data signal. The driver produces modulating forward current with a suitable amplification factor to produce the necessary output levels. In addition, the driver incorporates integrated speedup capabilities aimed at improving the turn-on and turn-off time of the optical source. These can be implemented using the bias circuit, charge injection circuit, and/ or charge extraction circuits. The bias circuit supplies a small quiescent current to the LED/LD at all times and, thus, keeping the space-charge capacitance of the source constantly charged. This effectively reduces the delay of carrier injection into the active region, minimizing delay in light output [24]. Charge injection works by introducing a large current spike at each positive edge of the data signal, which decreases the optical rise time of the emitter. Similarly, charge extraction improves optical fall time by sweeping out carriers from the active region of the emitter by giving the negative current spike at each falling edge of the data signal [25]. Few of the high-speed driver circuits used in the transmitter circuit are shown in Fig. 8.6.

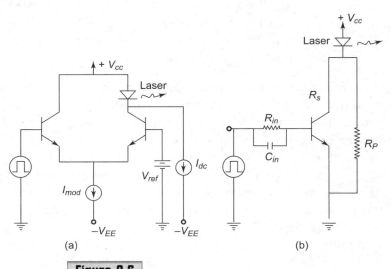

Figure 8.6　High-Speed Driver Circuits.

8.4.1.3 Modulators

The information signals can either directly modulate the laser light or an external modulator is used for modulation of the light generated by the laser source. In the case of direct intensity modulation, the source itself is modified by the information to produce a modulated output optical signal. The output optical intensity of the laser source changes in proportion to the injected current. For digital modulation the diode is modulated by a current source, driving the light source ON and OFF by the high-speed driver circuit switching. In the case of analog signals a laser DC bias is given more than the threshold value to keep the total current in the forward direction at all times and not letting the diode shut off in the negative swing of the signal current [26]. Linearization circuits are also used for analog modulators with negative feedback to reduce distortion [27].

When the data rate is very high internal modulation causes dispersion penalties due to frequency shift in the optical carrier leading to optical chirp. Therefore, external modulators are used at high data rates. In external modulator the light generation and modulation process are separated to avoid chirp and turn-on transient effects. The propagation characteristic of the modulator is altered by the electrical modulating signal and the optical light is then focused on this particular region of the modulator. Such systems have the advantage of utilizing the full power capability of the source. The commonly used modulators in optical communication are the electro-absorption modulator (EAM) and the Mach-Zehnder modulator (MZM). EAM's are usually wide-band, but they have higher loss (5-10 dB), poor extinction ratio (11-13 dB) and chirp of the order of 1dB. The monolithic EAM and CW laser source are compact with low coupling loss between the two devices. MZM on the other hand though much larger in size but have dynamic extinction ratio in the range of 15 to 17 dB and chirp of less than 0.1 dB [25]. Figure 8.7 shows the EAM and MZM external modulators.

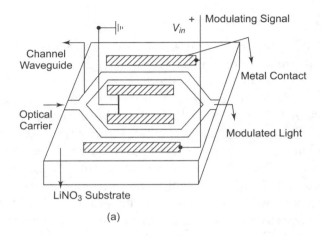

(a)

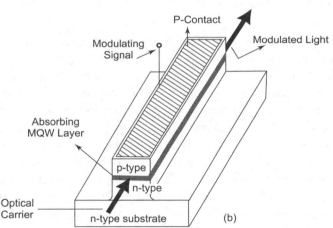

Figure 8.7 External Modulators. (a) Mach-Zehnder Interperometer Modulator (b) Electro-Absorption semiconductor Modulator (EAM)

8.4.2 Receivers

A typical optical wireless receiver or the front end consists of an optical system which collects and concentrates incoming radiation, an optical filter to reject ambient illumination, and a photo detector to convert the optical signal to photocurrent. Amplification, filtering and data recovery of the signal is then required to extract the transmitted signal. Figure 8.8 gives the block diagram of a FSO receiver. The light received from the atmosphere is incident on the collecting optics of the receiver, filtered and then detected by the receiver and converted to the electrical current. The electrical signal is then decoded and signal processing carried out to obtain the transmitted signal.

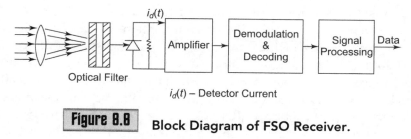

$i_d(t)$ – Detector Current

Figure 8.8 Block Diagram of FSO Receiver.

8.4.2.1 Receiver Optical Systems

The receiver optical system in the front end collects the transmitted radiation and focuses the incident light on the detector's active area. Both imaging and non-imaging optics [28] can be used to collect and focus radiation onto single element detectors. These are characterized in terms of their field of view angle θ_{FOV}, and the collection area. For constant radiance these are related to the detection area by the constant radiance theorem, which states that:

$$A_{ap} \sin^2\left(\frac{\theta_{FOV}}{2}\right) \le A_{det} \tag{8.7}$$

where, A_{ap} is the aperture or the collection area of the antenna and A_{det} is the photodectetor area. This is important as it limits the collection area that is available for a given FOV and photodetector. There are some well designed optical antennas, which have good performance in a compact form.

In the case of space diversity systems, a number of single channel receivers are combined and the antennae of each faces in a different direction. This allows multipaths to be resolved and collection areas for each to be increased [29]. The same function can also be provided by the imaging receivers [30] as well. A large area pixelled detector array is used to make the optical imaging system. Light from narrow range of directions is collected and resolved and imaged to different pixels on the array. As the array allows the large detection area to be segmented, it also reduces the capacitance on each of the receiver front ends and hence increases the bandwidth.

8.4.2.2 Optical Filtering

Optical filters in the front end are used to reject out of band ambient radiation, which greatly deteriorates the FSO link performance in the outdoor systems. Various different filter types have been demonstrated for this purpose; a long pass filter in combination with a silicon detector provides narrowing of the bandwidth [31], and absorption filters can be used to reject solar and other unwanted illumination. Bandpass interference filters are also used, although care has to be taken to allow sufficient bandwidth for passband shifting with the varying angle of incidence. Thin-film optical filters [32] are normally used in terrestrial links, which are constructed from multiple layers of thin dielectric film. It is also possible to filter by incorporating appropriate layers in the photodetector itself. Holographic receiver front-ends also allow ambient light noise to be rejected [33]. The constant ambient illumination which generates a DC photocurrent can be blocked by the AC coupling of the receiver itself.

8.4.2.3 Optical Detectors

The optical detectors used in FSO are *square law detectors*. This means that the output is proportional to the square of the incident optical field and hence has no direct dependence on optical phase, frequency, or polarization. Optical detectors can have very high speed, with bandwidths from DC to more than 50 GHz and they respond to the time average square of the field. The performance characteristrics for the detectors are high optical and power efficiency, reliability and ability to perform well in the presence of background noise and other atmospheric channel effects. In order to have increased channel bandwidth and reduced ambient illumination effect, the detector is to have a low capacitance and a narrow field of view. When the receiver collection aperture is smaller than the transmitter beam at the receiver, the power collected will usually be proportional to the detector area only. However, increasing detector area increases the capacitance presented to the system preamplifiers, which reduces the available bandwidth. This capacitance can be reduced by increasing the junction width but on the expense of the transit time and hence the speed. Therefore, an optimum detector for optical wireless should balance transit time and capacitance controlled bandwidth and achieve maximum area for a given bandwidth.

There are two basic detectors used in the optical systems: the *p-Intrinsic-n* (PIN) diodes and the Avalanche Photo Diode (APD). PIN receivers are commonly used due to their lower cost, tolerance to wide temperature fluctuations and operation with an inexpensive low-bias voltage power supply. But, the PIN receivers are about 10 to 15 dB less sensitive than APD receivers. The reduced sensitivity of the PIN receivers can be compensated to certain extent by increasing the transmitter power and using larger receiver lens diameter. In general, the high-rate optical communication systems utilize PIN photodiodes which are commercially available with electrical bandwidths exceeding 50 GHz. These detectors are many a time used as a close-packed array on a single chip and are illuminated through their substrate. Epitaxially grown structures allow independent control of depletion width by the growth of intrinsic regions of the correct thickness. Optimized detectors might also include filtering layers to reduce the noise from optically broadband ambient illumination. Devices thus fabricated can typically have capacitance of the order of 10 pF at 4-V reverse bias which is a factor of five lower than typical commercial devices [34].

APDs have internal gain that can be used to improve receiver sensitivity, although this comes at the expense of bandwidth and limited dynamic range of operation. On the other hand, with the increased power margin a more robust communication link can be built and this also reduces the criticality of accurate alignment of the link. However, the APD detector receivers are costly and need high operating voltages.

Detector Material   The two most common materials used for the detection of light in the near infrared spectral range are silicon and InGaAs. Figure 8.9 gives the response of the two materials with wavelength. Silicon is most commonly used detector material in the visible and near IR wavelength range. Silicon typically has the maximum sensitivity around 850 nm and has a very high bandwidth. Therefore silicon detectors are ideal candidates for light detection in conjunction with short wavelength 850 nm VCSEL laser.

Silicon drastically loses sensitivity toward the longer infrared spectrum for wavelengths beyond a micrometer. Operation of Si PIN/APDs at 10Gbps has been commercialized for use in shorter wavelength 850 nm for 10 Gb systems. Si-PIN detectors with integrated transimpedance amplifiers (TIA) are also used. Typical sensitivity values for a Si-PIN diode are around -34 dbm at 155 Mbps. Si-APD is far more sensitive due to an internal avalanche process. Sensitivity values for APD can be as low as -50 dbm at 10 Mbps. Therefore Si-APD detectors are very useful for lower light level detection in free space optics systems. In spite being very large in size (0.2x0.2 mm) silicon detectors still operate at higher bandwidths. This feature minimizes loss when light is focused on the detector by using either a larger diameter lens or a reflective parabolic mirror.

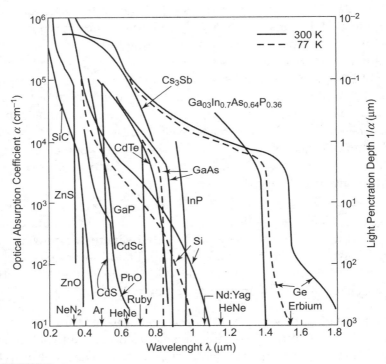

Figure 8.9 Spectral Response of Different Photodetectors (Emission Wavelength of Various Lasers are also Indicated) [35, page 665].

For 1550 nm wavelength radiation, InGaAs is the most commonly used detector material. Due to the drastic decrease in sensitivity towards the shorter wavelength range, InGaAs detectors are typically not used in the 850 nm wavelength range. The main benefit of InGaAs detectors is higher bandwidth capability. The majority of InGaAs receivers are based on PIN or PIN-TIA technology. Typical sensitivity values for InGaAs PIN diodes are similar to those of Si-PIN diodes. InGaAs diodes operating at higher speed are typically smaller in size than Si-PIN diodes because they are designed to be used in the single mode fiber system which does not require a large detection surface. This makes the light coupling process more difficult and overall losses increase when the light is coupled from free space onto the detector surface.

8.4.2.4 Receiver Types

Receiver structures used in optical systems can be either coherent or non-coherent (i.e., the direct detection receiver) types. Similar to RF coherent receivers optical coherent receiver are more sensitive but complex. They are frequently used in space communication, but in terrestrial links it is the direct detection receivers, which are mostly used because of their less complex and low-cost design. Figure 8.10 gives the schematic of a single PIN front-end receiver, structure. The pre-amplifier, which precedes the detector in most cases, is the trans-impedance design with a gain of G, load resistance R_L and noise figure of F_N. An electrical filter with a bandwidth of B_e is used for equalization with a impulse response inverse of the combined channel and preamplifier response. A sampler and threshold detector are used as in other electrical receiver to correctly estimate the transmitted signal.

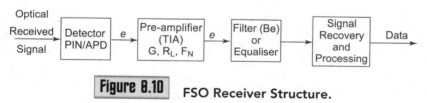

Figure 8.10 FSO Receiver Structure.

The direct detection receivers used in terrestrial links are in general Gaussian noise limited receivers with sufficient background ambient noise, though there are also Photon counting DD receivers which are Poisson noise limited and with much higher sensitivity.

The PIN photodiode-based detector structure is most common receiver for high-speed optical communications. While PIN-based receivers are relatively simple, they are less sensitive and require careful front-end electronic design and signal processing circuits to achieve good high-rate performance [36-37]. PIN detectors are also used in array receivers. Besides optimizing the detectors for low capacitance per unit area by increasing the width of the intrinsic region, input capacitance can also be mitigated by bootstrapping, equalization and capacitance tolerant front-ends [38-40]. PIN-based receivers offer superior bandwidth and dynamic range of operation, and are less temperature dependent offering robust performance over a wide range of environmental conditions. APD receivers though more sensitive, but are noisier and more temperature sensitive.

Pre-Amplifier The detector and the preamplifier, commonly known as the front-end, are the main determining factor in the performance of the receiver. The function of the preamplifier is to convert the detector current to a voltage signal with minimum addition of noise. The design of the front end requires a trade-off between sensitivity and bandwidth. The pre-amplifier can be of three categories; low-input impedance amplifier which has large bandwidth but high noise as well, high input impedance amplifier which though adds low noise, but also has lower speed, and the trans-impedance amplifier. In most cases, the trans-impedance design is used as it has a higher sensitivity together with a larger bandwidth. The dynamic range is also high for the trans-impedance amplifier. A typical trans-impedance amplifier configuration is shown in Fig. 8.11. This approach provides the best compromise in terms of noise, gain, and bandwidth, all of which are influenced by the capacitance of the detector.

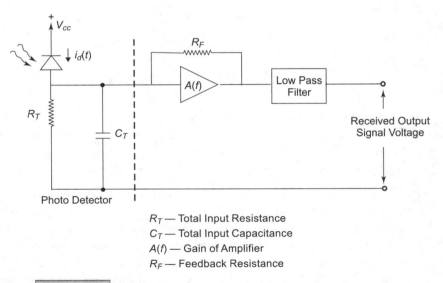

R_T — Total Input Resistance
C_T — Total Input Capacitance
$A(f)$ — Gain of Amplifier
R_F — Feedback Resistance

Figure 8.11   **Trans-impedance Pre-amplifier Configuration.**

The trans-impedance bandwidth is governed by the input capacitance, open loop gain of the amplifier, and the feedback resistance. With a total input capacitance, including detector and parasitic of 5 pF, the amplifier can achieve bandwidth up to 500 MHz. To increase the bandwidth to gigabit the receiver can be made by IC technology in a 0.35-μm CMOS process using combination of common-source channel stages and current-mode common-gate input stages [40-41].

8.5 NOISE IN FSO RECEIVER

The noise sources in FSO receiver are of two types: signal dependent and signal independent. The signal dependent noise results from shot noise due to random generation of electrons by the incident optical power. The shot noise is from both, the light signal and the background illumination light. It is also due to the dark current, which is present in the absence of any kind of light on the detector. The signal independent noise is the thermal noise due to the receiver itself and is scaled by the amplifier noise figure. The total power of noise sources of the FSO receiver shown in Fig. 8.10 with APD detector is expressed as:

$$\sigma_{sh}^2 + \sigma_{th}^2 = 2qF(M)(\Re P_s + i_d)M^2 B_e + \left(\frac{4kT}{R_L}\right)F_N B_e \qquad (8.8)$$

where, P_s is the incident signal power, $\Re$ the responsivity of the detector with $M = 1$ and i_d is the dark current. $F(M)$ is the excess noise factor due to multiplication factor M of the APD, which indicates the internal gain of APD. The typical values of M range from 3 to 100 and that of $F(M)$ from 2 to 10. In the case of PIN, these value reduce to 1. As the dark current noise is of the order of few nanoamps at 1550nm it can be neglected in (8.8). The thermal noise can be reduced by increasing the load resistance by using transimpedance amplifiers but the bandwidth of the system is limited due to residual capacitance. Commercial 1-kohm TIAs are available with $B_e = 7$ GHz suitable for 10 Gbps communications with typical noise figure of around 5 dB. For these parameters, the 10^{-9} BER can be achieved with −29 dBm power incident on the detector [42-43]. Another means of improving the performance of direct-detection receivers limited by thermal noise is to use low-duty-cycle return-to-zero (RZ) coding. The receiver performance is ultimately limited by choice of modulation format and coding.

Summary

The performance of a FSO link depends on the transmitter power, receiver sensitivity, receive aperture size, beam divergence, receiver and transmitter optics, bit rate and link range. The link and system design of FSO terrestrial communication system requires considering the issues regarding the link budget which will decide on the link margin required to obtain the desired performance from the designed link. Reliability of the link is an important criterion in any communication link. This can be determined from the availability of the link with the given BER or SNR performance for the given length of the link. The link power margins of most vendor equipment for FSO allow for availabilities that exceed 99.99% for the specified BER and range. Along with reliability of the link, component reliability is very important for the overall reliability of the system in the case of FSO links. As FSO is a wireless technology the maximum power levels transmitted from the sources are important consideration due to eye safety regulations. The power levels are limited by the IEC standards specified for acceptable laser radiations. The FSO systems use the same transmission wavelengths as that for the fiber based system. In the short wavelength region around 850 nm and in the long wavelength region around 1550 nm is used, as they fortunately are also the low attenuation band for the atmospheric channel. That makes readily available off-the-shelf commercial class transmitter and receiver modules used for the fiber systems with few changes required for the transmitter and receiver optics to couple light to/from the atmospheric channel, respectively. The design considerations of the various blocks in these modules are discussed. Also discussed are the different components of the noise which are added in the receiver. They have an impact on the receiver sensitivity which affects the link performance and range. Besides the transceivers, FSO transmission being based on LOS technology, require a robust mount with reasonably good automatic tracking and positioning system. This has to be an integral part of the transceivers at the two ends of the link.

References

1. Killinger D., 'Free Space Optics for Communication Systems', *IEEE Comm Mag*, pp. 36–42, Oct. 1989.
2. Arun K. Majumdar, 'Free-space laser communication performance in the atmospheric channel', *J. Opt. and Fiber. Commun.* Research, Vol.2, No. 4, pp. 345–396, 2005, Springer: New York.

3. Kim I., McArthur B., Korevaar E., 'Comparison of laser beam propagation at 785 nm and 1550 nm in fog and haze for optical wireless communications', Proceedings SPIE, 4214, 26–37 (2001)

4. David F., Giggenbach D., et al., 'Preliminary results of a 61 km ground-to-ground optical IM/DD data transmission experiment', Proceedings SPIE, Vol. 4635 (2002)

5. Leitgeb E., Gebhart M., and Birnbacher U., 'Optical networks, last mile access and applications', *J. Opt. and Optical Fiber Communication Research*, 56–85 (2005) Springer: New York.

6. Gebhart M., et al., 'Measurements of light attenuation at different wavelengths in dense fog conditions for FSO applications', Report of STSM (COST 270), June 2004.

7. Sheikh Muhammad S., Köhldorfer P., Leitgeb E., 'Channel modeling for terrestrial FSO Links', Proc. ICTON 05, July 2005, Barcelona, Spain.

8. Eric Korevaar, Isaac I. Kim and Bruce McArthur, 'Atmospheric propagation characteristics of highest importance to commercial Free Space Optics', Proceeding SPIE, pp 1–12, Vol. 4976, (2003).

9. www.lightpointe.com

10. 'Generic reliability assurance requirements for optoelectronic devices used in telecommunications equipment', Telcordia GR-468-CORE no. 2, Dec. 2002.

11. 'Generic requirements for fiber optic branching components,' Bellcore GR-1209-CORE, Issue 1, Nov. 1994.

12. 'Generic reliability assurance requirements for fiber optic branching components,' GR-1221-CORE, Issue 1, Dec. 1994.

13. Lightpointe-White Paper series 2009, *How to Design a Reliable FSO System* Lightpointe.com/whitepapers/LPC FSO SystemDesign.pdf (14th July 2010)

14. Occupational Safety & Health Administration (OSHA), U.S. Department of Labor, 'Technical Manual on Laser Hazards', U.S. Occupational Safety and Health Administration legal standards, *www.osha.gov/dts/osta/otm/otm_iii_6.html*.

15. International Electrotechnical Commission, *Origin of International Laser Safety Standards:* IEC 60825–1 and IEC-60825-2, IEC, Geneva.

16. O'Brien D.C., et al.: Short range optical wireless communications, *Wireless World Research Forum*, 1–22, 2005.

17. Gfeller F.R. and Bapst U., 'Wireless in-house communication via diffuse infrared radiation', Proceedings of the IEEE, 67(11): pp. 1474–1486, November 1979.

18. Dominic C. O'Brien, et. al., 'Integrated transceivers for optical wireless communications', *IEEE Journal of Selected Topics in Quantum Electronics*, Vol. 11, no. 1, Jan/Feb 2005, pp 173–183.

19. Stach M. et al., 'Bidirectional optical interconnection at 100 Gbps data rates with monolithically integrated VCSEL-MSM transceiver chips', *IEEE Photonic Technology Letters*, 18(22), 2286–2388, 2006.

20. Senior J.M., *Optical Fiber Communications: Principles and Practice*, 2nd ed. Prentice Hall International, 1992.

21. Shinagawa T., 'Detailed investigation on reliability of wavelength-monitor-integrated fixed and tunable DFB laser diode modules', *J. Lightwave Technol.* **23**, pp. 1126–1136 (2005).

22. Agrawal G.P., *Fiber-Optic Communication Systems* (NewYork: JohnWiley & Sons, 1992).

23. David O. Caplan, 'Laser communication transmitter and receiver design', *J. Opt. and Fiber. Commun. Research*, pp. 225–362 (2007) Springer New York.

24. Coldren L.A. and Holburn S.D.M., Lalithambika V.A., Samsudin R.J., Joyner V.M., Mears R.J., Bellon J., and Sibley M.J., 'Integrated CMOS transmitter driver and diversity receiver for indoor wireless links,' *Proc. SPIE, Optical Wireless Communications V*, Vol. 4873, pp. 13–21, 2002.

25. Corzine W., *Diode Lasers and Photonic Integrated Circuits*, JohnWiley & Sons, Inc., 1995.

26. You R. and Kahn J.M., 'Average Power Reduction Techniques for Multiple-Subcarrier Intensity-Modulated Optical Signals', *IEEE Trans. on Commun.*, Vol. 49, pp. 2164-2171, December 2001

27. Roberto Remirez-Iniguez, Sevia M. Idrus & Ziran Sun, 'Optical Wireless Communications': *IR for Wireless Connectiviy'*, CRC Press, 2008.

28. Welford, W.T., *The Optics of Nonimaging Concentrators: Light and Solar Energy*. 1978, New York: Academic Press.

29. Carruthers, J.B. and J.M. Kahn, 'Angle diversity for non-directed wireless infrared communication', *IEEE Transactions on Communications*, 2000. **48**(6): pp. 960-969.

30. Kahn, J.M., et al., 'Imaging diversity receivers for high-speed infrared wireless communication', *IEEE Communications Magazine*, **36**(12): p. 88, 1998

31. Kahn J.M. and Barry J.R., 'Wireless Infrared Communications,' *Proceedings of the IEEE*, Vol. 85, no. 2, February 1997, pp. 265–298.

32. Keiser G., *Optical Fiber Communication*, 4th edition, Tata-McGrawHill Publications, New Delhi, India (2008).

33. Jivkova, S. and Kavehrad M., 'Holographic optical receiver front end for wireless infrared indoor communications', *Applied Optics*, 2001. **40**(17): pp. 2828–35.

34. O'Brien D.C., et.al., 'Integrated transceivers for optical wireless communications', *IEEE Journal of Selected Topics in Quantum Electronics*, Vol. 11, no. 1, Jan./Feb. 173–183, 2005.

35. Yariv A. and Yeh P., 'Photonics: Optical Electronics in Modern Communications', 6th Ed. Oxford University Press-NY, 2007.

36. Personick S.D., 'Receiver design for digital fiber optic communication systems, I & II', *Bell Syst. Tech. J.* **52**, pp. 843–886, 1973

37. Smith R.G. and Personick S.D., 'Receiver Design,' in *Semiconductor Devices for Optical Communication*, H. Kressel, Ed. (NewYork: Springer-Verlag, 1980)

38. Alexander S.B., *Optical Communication Receiver Designs*, SPIE Optical Engineering Press, Bellingham, Washington, USA: (1997).

39. Green R.J. and McNeill M.G., 'Bootstrap transimpedance amplifier: A new configuration', *IEE Proceedings*, 136, 57–61, (1989).

40. Chia-Ming Tsai, 'A 40 mW 3 Gb/s self-compensated differential transimpedance amplifier with enlarged input capacitance tolerance in 0.18 μm CMOS technology,' *IEEE J. of Solid-State Circuits*, Vol. 44, no. 10, Oct. 2009, pp. 2671-2677.

41. Sackinger E., *Broadband Circuits for Optical Fiber Communication*, J. Wiley & Sons, 2005

42. Miyamoto Y., Hagimoto Y., and Kagawa T., 'A 10 Gb/s high sensitivity optical receiver using an InGaAs-InAlAs superlattice APD at 1.3 μm/1.5 μm,' *IEEE Photon. Technol. Lett.* **3**, pp. 372–374 (1991).

43. Yun T.Y., Park M.S., Han J.H., Watanabe I., and Makita K., 'A 10-gigabit-per-second high-sensitivity and wide-dynamic-range APD-HEMT Optical Receiver', Photonics Technol. Lett. **8**, 1232–1234 (1996).

Index

A

Absorption 18
Absorption cross section 19
Aerosol scattering OTF 53
Alamouti scheme 110
Angular spread 17
Antenna averaging 37
Antenna diversity 107
APD 216
a-posteriori 164
a-priori probability 164
â-priori probability 72
Array gain 108
Atmospheric losses 18
Autocorrelation function 74
AWGN 42

B

Bandwidth efficiency 67
Bandwidth efficient schemes 68
Bandwidth expansion ratio 165
Bandwidth-limited regime 133
Basis function 70, 75
Bayes's rule 72
Beam divergence 17
Beam spreading 27
Beam wander 27
Beer's Law 18
Belief propagation 185
BER 106
Best resource allocation 140
Bias circuit 212
Bipartite graphs 183

Bit Error Rate 71
Block codes 162
Bootstrapping 219
Bose-Chaudhuri Hocquenghem (BCH) 167
Building sway 29, 202

C

Ceiling-bounce model 59
Channel state information 114
Charge extraction 212
Charge injection 212
Chip rate 78
Chips 78
Chirp 213
Code gain 163
Code rate 165
Code words 165
Codebook 165
Coherence bandwidth 43, 44
Coherence time 43
Coherent 219
Complementary cumulative distribution function (CCDF) 100, 137
Constellation 73
Constraints 38
Continuous spectrum 74
Cornea 204
Correlation length 48
Cycle 183
Cyclic Codes 166

D

Dark current 221
Decoder 162

Degree distribution 184
Degrees of freedom 108, 117
Delay spread 45, 59
Detection complexity 109
Differential Phase-Shift Keying (DPSK) 92
Differential PPM (DPPM) 90
Diffused channels 58
Digital Pulse Interval Modulation (DPIM) 92
Dirac delta functions 74
Direct detection 219
Discrete 74
Distributed Feedback Laser (DFB) 203
Divergence angle 17
Diversity gain 108
Diversity techniques 97

E

Eigenvalue decomposition 148
Electro-absorption modulator (EAM) 213
Encoder 162
Enumerator 178
Equal gain combining 114
Equalization 219
Ergodic capacity 135
Error floor 177
Euclidean distance 72, 163
Excess noise factor 221
External modulators 213
Eye safety 67
Eye-safe 204

F

Fabry-Perot (FP) 203

Fading 36
Fast fading 44
Feedback resistance 220
Flat fading 45
Forward Error Control (FEC) 160
Frequency-selective 45

G

Gamma-Gamma 49
Gaussian 50
Gaussian elimination 184
Gaussian model 102
Generator matrix 163
Generator polynomial 166
Girth 183

H

Hamming distance 163, 165
Hamming weight 163
Hard-decision 163
High input impedance amplifier 220
Holographic receiver front-ends 216
Hybrid Concatenated Convolution 175
Hybrid non-LOS 58
Hybrid Pulse Amplitude and Position Modulation (M-n-PAPM) 92

I

I.I.D 70
Ideal flat channel 68
IEC standard 204
Image dancing 27
Image plane 41
Impulse response 40

InGaAs 217
InGaAs detectors 218
Inner and outer length scales 48
Inner scale 45
Integrated Transimpedance Amplifiers (TIA) 217
Interference filters 216
Interleaving 107, 161
Inter-symbol-interference 37
Irregular codes 183

L

Likelihood ratios 164
Linear 163
Log-Likelihood Ratio (LLR) 99, 164
Log-MAP 177
Log-normal 49
Log-normal distribution 26
Long pass filter 216
Low Density Parity Check (LDPC) codes 161
Low-input impedance amplifier 220

M

Mach-Zehnder Modulator (MZM) 213
Marginal distribution 50
Matched filters 71
Maximum â-posterior probability (MAP) 72
Maximum likelihood 72
Maximum-Likelihood Sequence Detection (MLSD) 77
Max-Log-MAP 177
Mean capacity 135

Mean-Time-to-Failure (MTTF) 203
Message Passing 185
Mie scattering 18, 22
MIMO 108
MISO 108
ML detector 72
MODTRAN 32
Modulation Transfer Function 41
Molecular and aerosol absorption coefficients 19
M-PPM 78
Multilevel 81
Multi-path 36
Multiplicative noise 134
Mutual Coherence Function 47

N

Negative exponential 49
Noise sources 221
Non-coherent 219
Nondirected-non-LOS 58
Non-linear 163
Non-LOS configuration 58
Non-negativity constraint 38
Non-selective scattering 18
Normalized covariance 48

O

Object plane 41
On-Off Keying (OOK) 75
Open loop gain 220
Optical Alamouti Scheme 110
Optical filters 216
Optical hardware OTF 53
Optical transfer function 41
Optimum detection 70

Optimum detector 71
Orthogonal signaling 78
Orthonormal functions 70
Outage capacity 136
Outer scale 45

P

Parallel Concatenated Convolution 175
Parity check matrix 163
Phase transfer function 41
Pilot symbols 108
PIN 216
Point spreading function 41
Poisson Photon-counting Model 103
Poisson point process 103
Power efficiency 67
Power Spectral Density (PSD) 74
Power-efficient modulation schemes 67
Power-limited regime 133
PPM 67
Pre-amplifier 219, 220
Probability of bit error 73
Probability of symbol error 73
Pulse Amplitude Modulation (M-PAM) 81

Q

QR decomposition interference suppression 189

R

Radial variance 27
Raleigh scattering 21

Rank deficient 148
Rayleigh scattering 18
Receive diversity 108
Recursive systematic convolution (RSC) 172
Redundant 162
Reed Solomon codes 161
Regular and irregular LDPC code 184
Reliability 203
Repetitive diversity MIMO 112
Retina 204
Rytov variance 26

S

Safety standards 204
Scattering 18
Scattering coefficients 19
Scintillation 26
Seismic activities 29
Sensitivity 217
Sequential decoding 173
Serial concatenated convolution 175
Shot noise 221
Signal constellation 70
Signal dependent 221
Signal dependent noise 221
Signal independent 221
Signal space 68
Silicon 217
Silicon detectors 217
SIMO 108
Singular value decomposition 148
SISO link 104
Skin-safe 204
Slow fading 44
Slow fading channel 135

SNR 100
Soft bit 164
Soft Output Viterbi Algorithm 177
Soft-decision 163
Space Diversity 107
Space-time coding 109
Spatial frequencies 41
Spatial multiplexed 112
Spatial multiplexing gain 109
Spectral broadening 43
Spectral efficiency 133
Sphere-packing 130
Square law detectors 216
State sequence 172
Structure coefficient 25
Structure parameter 46
Sum-Product 185
Symbol and chip synchronization 81
Symbol error rate (SER) 100
Symbols 162
Syndrome decoding 166
System OTF 53
Systematic 165

T

Table look-up decoding 173
Tanner graph 183
Telcordia 203
Temporal dispersion 37
Thermal noise 221
Thin-film optical filters 216
Threshold decoding 173
Threshold detection 77, 106
Time Diversity 107
Time selective fading 107
Time variations 36

Transfer function 41
Trans-impedance amplifier 220
Transmission windows 24
Transmit diversity 108
Tree codes 162
Trellis 172
Turbo codes 161
Turbulence 25
Turbulence eddies 26, 45
Turbulence OTF 53

U

Union bound 73
Unshadowed 60

V

Vector channel 68
Vertical Cavity Surface Emitting
    Lasers (VCSEL) 203
Viterbi Algorithm (VA) 173, 177

W

Water-filling 140
Wavelength Diversity 107
Weight enumerator function 178
Whitened Matched Filter 77
Wide-sense cyclostationary 74
Wiener-Kinchine theorem 74